QUALIFYING FOR ADMISSION TO THE SERVICE ACADEMIES

A Student's Guide

QUALIFYING FOR ADMISSION TO THE SERVICE ACADEMIES
A Student's Guide

by

Robert F. Collins, Col. (ret.)

THE ROSEN PUBLISHING GROUP, INC.
New York

Published in 1987, 1990 by The Rosen Publishing Group, Inc.
29 East 21st Street, New York, NY 10010

Revised Edition 1990

Library of Congress Cataloging-in-Publication Data

Collins, Robert F., 1938–
Qualifying for admission to the service academies

(The Military Opportunity series)
Includes index.
1. Military education—United States. 2. Naval education—United States. 3. Aeronautics, Military—Study and teaching—United States. I. Title.
II. Series
U408.C58 1987 355′.007′1173 87–9717
ISBN 0–8239–1187–X

Manufactured in the United States of America

To the outstanding young men and women who have been commissioned at the service academies and who have served their nation with courage, honor, dignity, integrity, and dedication.

About the Author

Robert F. Collins is a recently retired officer who served more than twenty-five years in the US Army. He enlisted in the Army in 1960, graduated from Officer Candidate School in 1962, and achieved the rank of Colonel before his retirement in 1985. Col. Collins was in the Military Intelligence branch and served tours of duty in Korea, Vietnam, Germany, and the United States. He is a Soviet Foreign Area Officer with extensive travel in Eastern Europe and the Soviet Union. He taught US National Security Policy at the US Army's Command and General Staff College for five years and was a Professor of Military Science for two years. His decorations include the Army Commendation Medal, the Meritorious Service Medal, the Bronze Star, and the Legion of Merit.

Acknowledgments

The author wishes to acknowledge the assistance provided by representatives of the Army, Naval, Air Force, Coast Guard, and Merchant Marine Academies. They were not only extremely helpful in providing source material for the original edition of this book, but they have provided updated material and recommendations for this revised edition. Much of the material in this book has been drawn directly from service publications to insure accuracy and completeness. Errors in the book, of course, are mine.

Contents

Introduction

In the Armed Forces of the United States, five service academies are maintained by the federal government: the United States Military Academy (USMA) at West Point, New York; the United States Naval Academy (USNA) at Annapolis, Maryland; the United States Air Force Academy (USAFA) at Colorado Springs, Colorado; the United States Coast Guard Academy (USCGA) at New London, Connecticut; and the United States Merchant Marine Academy (USMMA) at Kings Point, New York. Each academy operates under the general supervision and direction of its respective service except the Coast Guard Academy, which operates under the general direction and supervision of the Coast Guard, which is Part of the Department of Transportation; and the Merchant Marine Academy, which is operated by the Maritime Administration of the Department of Transportation.

The appeal of the service academies to the young men and women of America is remarkable. Generally speaking, there are more than ten applicants for every space available, and the ratio is much higher if all applicants for congressional appointments, even those turned down, are included. A renewed spirit of patriotism is evident among America's citizens after having reached a low point in the late sixties and early seventies during the Vietnam War. Obviously, many factors contribute to the current sense of pride and desire to serve the country. Current events in Eastern Europe and the Soviet Union serve to emphasize the appeal of democracy to all people, and all Americans are proud of the fact that for most people in the world, America is synonymous with democracy. The end result is that competition to enter the service academies will become even sharper in the coming years. It is also possible that the number of spaces in the academies, now established by law, could decrease as a part of the program to reduce defense expenditures. It is therefore all the more important that aspirants be fully knowledgeable about application procedures, academic standards, physical qualifications, and academy life.

The purpose of this book is to present in one volume information on the five service academies that will be helpful to young men and women who are considering applying to the academies and entering a subsequent military career or a career in the Coast Guard or the Merchant Marine. It is hoped the book will be a useful guide to high school students who are considering service careers, to their parents, to high school guidance counselors, to cadets currently enrolled in the academies, to education officers at bases, posts, and at sea, to ambitious young service personnel on active duty, and to the public at large. No attempt is made to minimize the difficulties of gaining admission to the academies or to soften the rigors of discipline required to complete the courses to graduation. The young men and women who succeed will be physically fit, morally sound, and socially adept. Most important, they will be strongly motivated to serve their country.

Potential applicants are advised to examine closely their reasons for wanting to attend a service academy. It is a demanding four years, and attrition rates are high. However, the rewards are great for newly commissioned officers. They have received one of the finest educations in the country at government expense, a bachelor of science degree with continuing education guaranteed, and a commission in one of the services with a promising career before them.

One of the main themes of the book is that you should obtain as much information as possible about each service academy before you make a decision to attend. You incur obligations when you attend the academies. Most of them have an eight-year obligation: at least five years on active duty and then a three-year stint in the inactive reserve. Generally speaking, you do not incur an obligation if you do not enter the third year at the academy. However, if you do enter the third year and do not successfully complete the four-year curriculum or do not enter into active duty, you may be subject to service in an enlisted status, or you may have to reimburse the government for educational expenses. It is important that you fully understand your obligations and commitments before signing any agreement with the government. It is also important to understand that each of the services may have special provisions and unique procedures. Learn as much as you can about your academy before making irrevocable decisions. Talk to friends who have attended the academy, talk to your guidance counselors, teachers, parents, friends, academy representatives. The academies are eager to provide information about their respective institutions. They will also encourage you to visit the academy. You can obtain information by writing the Admissions Office at all of the service academies. The addresses are given elsewhere in the book.

You should examine carefully your reasons for wanting to attend a service academy. It is not a decision to be made lightly. Going to the academy as a result of pressure from your friends or from your parents is not a good way to get started. Going to the academy to get away from problems at home is never a good course. The decision to attend a service academy must be yours and yours alone. If you are not sufficiently motivated, you will not complete the four difficult years successfully.

The words of President John F. Kennedy, spoken almost thirty years ago in August 1963, are as valid today as they were then. He said in part, "So I express our very best wishes to you and tell you that though you will be serving in the Navy in the days when most of those who hold public office have long gone from it, I can assure you in 1963 that your services are needed, that your opportunities are unlimited, and that if I were a young man in 1963 I can imagine no place to be better than right here at this Academy, or at West Point, or in the Air Force, or in some other place beginning a career of service to the United States."

Chapter I

Into the Twenty-first Century

Changes in the world situation are occurring at a dizzying pace. It is possible that historians will look back upon the last decade of this century and judge it to be one of the important turning points in modern times. At no time since the end of World War II have prospects for genuine reconciliation and peace looked so promising. The opening of borders between East and West, symbolized by the tearing down of the Berlin Wall, is a profound event with far-reaching implications and repercussions. The simultaneous movement of so many different peoples and nations to demand the basic liberties that we enjoy: freedom of expression, freedom to vote with a choice of candidates, freedom to disagree, freedom to form political parties, freedom from unreasonable search, and all the civil liberties essential to a democracy, is one more strong indication that communism is not a viable form of government. These monumental political changes have profound effects on the US defense structure and US defense policies. Important questions about the size of the US military, the amount of money spent for defense, the continued development of nuclear weapons, and even the *need* for a strong military posture are sure to be raised and debated in the next ten years.

In a time that many analysts are hailing as a new epoch in East-West relations, many groups will call for new and greater agreements on disarmament. Some groups will even argue for unilateral disarmament, while other factions will argue for maintaining strong defense forces as well as increased vigilance in what they see as unsettled times. To further exacerbate this contentious issue, ongoing advances in technology will provide an increasing number of nations with the capability not only to develop nuclear weapons, but to deliver them.

It is ironic that technology will be an argument that both sides will use to advance their own position. The side arguing for major disarmament will state that advances in technology will allow the US to perform worldwide surveillance so that it would be impossible for any nation to prepare for war without detection. Technology in the overhead surveillance vehicles will become so sophisticated that despite a nation's efforts to hide its weapons deployments or its technological breakthroughs, the US will have sufficient lead time to prepare for attack. Proponents of disarmament will also argue that technology has made such advances in the lethality of weapons that it is senseless to pursue confrontational policies with other nations. Continued support for strong military programs will be perceived as hostile and antagonistic by other nations and will result in continued military buildup by those nations. This, they argue, can result only in conflict that can mean the end of civilization because of the destructive power technology has provided many nations. According to this side, technology has brought nations closer together and shrunk the world. Nations no longer act in isolation; one nation's actions affect the entire international community. Consequently, we owe it to future generations to make every effort to preserve peace. Technological communications breakthroughs have given nations the ability to communicate on a real-time basis so that misunderstandings are less likely to occur. This side is likely to contend that the current warming of East-West relations provides an unprecedented opportunity for the US to take dramatic chances to improve world peace. If we show the rest of the world, especially the Soviet Union, that we are serious about peace and that we will cut back on our defense spending and military establishment, the other countries will follow suit. The US is an acknowledged world leader, and it is our responsibility and obligation to do whatever we can to improve the chances for peace, even if it means taking risks by relying on other nations' good intentions.

The side arguing for a continuing strong military posture will also use technology to persuade voters of the soundness of their position. There is no way of controlling, much less predicting, the possible outcomes of the technology explosion. Although it is easy to talk of being able to monitor other nations' activities, in point of fact a nation could achieve a technological breakthrough that could render all our surveillance technology obsolete. Such a development could occur so quickly that we would not be aware of it, or if aware of it, we would not be able to respond adequately. To extend the argument, technology theoretically has the potential to create a weapon powerful enough to destroy the US before the US could retaliate.

This side goes on to point out that our military strength is the main reason that East-West relations have improved. Once we demonstrated to the Soviet Union that we had the will and determination to protect our national interests and assist countries friendly to the US, the Soviet Union began to modify its aggressive policies and think seriously about improving relations with the US. The political changes occurring in the Eastern European countries, notably East Germany, Poland, Czechoslovakia, Romania, and Hungary, are natural outgrowths of the improving relations between the United States and the Soviet Union. If the United States had not been adamant about human rights for all peoples, domestic conditions would not have improved in the USSR; the Soviet leadership did allow *glasnost* (openness) and *perestroika* (restructuring) to occur within the country, and that led to demands for more freedoms in countries heavily influenced by the Soviet Union.

Proponents of a strong military structure into the twenty-first century point out that history has repeatedly demonstrated that nations tend to become more aggressive if they perceive that other nations are disarming. Peace is a desirable goal without question, but we must be cautious and not undo all we have accomplished in the past ten years by acting precipitously and disarming too hastily. In their defense philosophy, peace through strength is their credo.

The significance of this basic argument for future military

officers of the United States is that, barring a national emergency or some catastrophic turn of events, defense dollars will have to be fully justified on a continuing basis. There will always be a demand for more dollars than are available, and defense will have to set its priorities just as will other programs. The United States will face staggering problems relating to drugs, the environment, aging, the economy, health care, nuclear power and waste, Third World poverty, and rising expectations of the inhabitants of less developed countries. Added to this list are concerns about dwindling resources, world overpopulation, the Middle East dilemma, international terrorism, and the ongoing struggle for independence in many of the world's nations. The US military must be seen against this backdrop of problems to be confronted in the 1990s. The military does not operate in isolation; its policies and direction are provided by duly elected civilian governments, and many decisions made by our government have direct effects on funding for the military as well as use of the military. Let's examine briefly some of the problems that have to be faced as we move into the twenty-first century.

In times of relative calm in international relations, the major concerns of US citizens center on domestic issues and social changes. This is probably the case for the next ten years; barring unforeseen events, most Americans will have a plethora of domestic and social items to resolve for the immediate future. However, it is important to understand that some problems are so complex and intractable that they will still defy resolution. Americans generally are optimistic, believing that all problems can be resolved if we work hard enough and take a reasoned approach. Unfortunately, we are coming to realize that some problems are impossible to resolve. An example may be the sale and use of drugs in the United States, which has so many facets that it seems overpowering.

All indicators show that the problem can only get bigger and more complex in the future. The billions of dollars generated by drug trafficking has never been accurately estimated. The human cost in death and misery caused by drugs cannot be estimated, much less the potential value of the contributions lost to society by young people ruined by drugs.

Consider some of the many other problems: the continuing battle over appropriate penalties for drug sellers and users, the crowding of court dockets with drug cases, the overcrowding of prisons and detention centers, the increase in violent crimes, health problems associated with drug users and their children, the siphoning off of increasing numbers of law-enforcement officials to deal with the problem, the alienation of young people from society because of drugs, the loss of billions of dollars of tax revenues in illicit drug trading, the strain in relations with large drug-producing countries, and the long-term medical problems associated with drug use. The cost of treatment and rehabilitation increases daily. Many differing solutions are proposed to these complex problems. Plans as extreme as the death sentence for drug dealers and legalizing of drugs are put forward. The fact is that no solution is available at present; and as long as staggering sums of money are involved in the drug trade and Americans use more and more forms of drugs, there will be no solution.

The drug problem affects the military in several ways. Indirectly, resources devoted to fighting the drug war will never be available to the military; directly, the military is likely to become more actively involved in the drug problem. The military could participate in surveillance, guarding borders, making drug sweeps, maintaining files, providing drug education classes, and even attacking drug strongholds if some proposals are adopted. No matter what proposals are enacted or what new ideas appear to solve the problem, drugs will be a social problem well into the twenty-first century.

Another full-time problem will be managing, controlling, or even merely predicting the turns of the US economy. Our economy is becoming more intertwined with those of other nations. The United States is well aware of how dependent its economy is on international capital flows and foreign trade accounts. More and more the economies of the modern democratic nations are linked together so that they share in troubled times and in prosperity. A recession or stock market crash in Japan or West Germany could mean an increase in interest rates in the US. No country in the world is even close to being self-sufficient. One has only to recall the Arab oil embargo in the early 1970s to understand how crucial foreign suppliers are to our economic well-being. It is likely that the US economy in the next decade will become even more intertwined with those of other nations and balance-of-trade payments and deficit spending will be a constant concern.

One of the vexing questions for the US has always been how far to encourage trade with the Soviet Union. Political considerations have carried greater weight than strictly economic considerations in our dealings with the Soviet Union over the past forty-five years. American allies have been far more willing to do business with the Soviet Union. However, that situation may change dramatically in the next decade. The United States can use preferential economic treatment toward the Soviet Union as one more incentive to maintain the programs started under *glasnost.* The US has been unwilling to trade freely with the Soviet Union because that would promote the stability of a regime that was actively opposing US interests in other parts of the world. The US has also been unwilling to supply technological items—computers, guidance systems, aircraft—lest they be used to modernize the Soviet Union's defense forces. It appears that as the Soviet Union moves toward more cooperation with the US, trade will increase and closer economic relations will result.

No one can predict with certainty what will occur in the US economy during the next ten years. Most analysts believe it is basically sound, although those same analysts are quick to point out the many variables and unknowns that make it hard to predict with confidence. The US economy is the largest in the world, and some people say it is too complex to comprehend. The bad news is that the economy can only become more complex; it will never simplify. Coupled with the known demands on the economy, new massive demands can arise unexpectedly. A case in point is the tremendous amount of money required for research, treatment, prevention, and legal costs associated with AIDS (acquired immunodeficiency syndrome). Twenty years ago no one had even heard of AIDS. Today and into the predictable future, expenditures for AIDS-related projects will be massive.

That is an example of a new demand upon limited resources that will probably increase annually. It is possible that other new demands for dollars will arise in the 1990s. For the military professional, the battle for allocation of resources will probably become even more competitive. As tensions ease, demands will be heard to reallocate defense dollars to domestic social needs: housing, medical care, education, research, highway construction, mass transportation, administration of justice, crime prevention. Historically, Americans have to be completely convinced of a threat to their security before they allocate adequate funds for the defense establishment. With all the domestic demands, it is likely that the defense establishment will decrease in size. That means that the size of the active duty forces will decrease, the force structure and associated equipment will decrease, and the United States will probably scale down its overseas commitments. Under those circumstances, the military will continue to need dedicated, well-qualified professionals to provide leadership.

Environmental concerns will increase greatly in the next decade along with the amount of money allocated to improve the environment. Under that broad heading fall a number of thorny issues: ecological concerns, air quality, the greenhouse effect, disposal of nuclear wastes, nuclear power, dwindling natural resources, overpopulation, and many other matters that affect the quality of life.

If there is one area in modern life that requires cooperation of all nations, it is the environmental problems. About eighty million people are added to the world's population each year, primarily in areas that are least able to accommodate them. The continuing rise in population of less developed countries increases the destruction of agricultural lands and water resources as well as frustrating efforts to combat world poverty. The ever-expanding desire for more material goods necessarily means more pollution. Until recent times we have not been concerned about the environment, being content to dispose of our wastes in the ocean or the atmosphere. We have damaged the ozone layer with industrial chemicals. That presents not only a potential increase of skin cancer because of the increase in ultraviolet radiation reaching the earth, but also a deleterious effect on crops. Increasing use of fossil fuels plus wholesale destruction of rain forests, especially in Brazil, releases greater quantities of carbon dioxide into the air, reducing humidification and the production of oxygen. The overutilization of ocean fisheries is calamitous.

These and other environmental problems will not disappear. There are encouraging signs that governments are becoming aware of the magnitude of the problem. Even to attack the problem, it is imperative that all nations cooperate. Our resources are not infinite, and our environmental balance is fragile. It is to be hoped that shared interest in protecting the environment will cause all nations to work together and decrease tensions in other areas.

It will not be an easy walk into the twenty-first century. We have mentioned only a few of the problems to be dealt with. Fortunately relations between the US and the Soviet Union seem to be improving. The Communist bloc in Eastern Europe appears to be rapidly dissolving, which may lead to more peaceful conditions in all of Europe. Grounds for hope exist in improving our relations with nations in other parts of the world. Our country will always need a military to safeguard our interests; if we have a continuing period of relative peace, the overall size of the military along with our commitments overseas will decrease. Barring a national emergency of some sort, there will be no reinstatement of the draft. The military will be all voluntary, and all services will be looking for talented, highly motivated young men and women. Competition for available slots in the service academies will probably become even more intense. Incentives to attract prospective candidates to the academies will probably increase. Attendance at an academy and a subsequent career in the military will remain a rewarding and exciting life.

Chapter II

Today's Professional Officer

The officer in today's armed forces must understand the unique role that the military plays in the American form of government. It is a role that has been shaped by experience, tradition, and the American values of individual worth and personal freedom. The tradition of military service has never developed in the United States; indeed, from the beginning of the American national experience it has generally been agreed that a large standing army was neither desired nor required. That belief was based on many circumstances. Starting with the colonial experience, with hope for the future and tremendous optimism the settlers were strongly independent and convinced that they were able to protect themselves. One of the reasons they had left Europe was to escape militarized, regimented, authoritarian societies; the American experience encouraged self-reliance, with security concerns best handled by oneself and one's neighbors. It is only since the end of World War II that Americans have come to realize the importance of having relatively large professional armed forces not only in place, but also prepared to fight on short notice.

The United States has been blessed by its geographical location. Until the beginning of the twentieth century it was relatively invulnerable to attack by foreign powers. Flanked by oceans on the east and west and having friendly neighbors on the north and south, the United States developed its traditions and its view of the rest of the world in a secure, relatively isolated manner. An abundance of natural resources, temperate climate, and productive agricultural lands further promoted its independence and self-reliance. The United States did not play a world role until this century, and no large standing military force was required to protect it from invaders, keep the sea-lanes open, or guarantee the freedom of the air. When faced with an emergency, the American people rallied to the call for arms, fought bravely, and attempted to resolve the conflict quickly. Military leaders have emerged in times of crisis and received honor and admiration for their deeds, but the American people have always insisted that the citizen army be disbanded as soon as the crisis was over.

Americans have always been sensitive to the dangers of too strong a military influence on government. Indeed, one of the basic tenets of our democracy, guaranteed by the Constitution, is civilian control of the military. It is an inviolable rule that the military only carries out policy; its role is to advise civilian decision-makers and then implement their decisions. Viewing events in eighteenth- and nineteenth-century Europe and military takeovers by Oliver Cromwell, Frederick the Great, and Napoleon Bonaparte reinforced the American idea of a small military to be expanded with citizens when emergencies arose. This approach worked well for the United States until this century, but now circumstances and the world situation have altered drastically.

The United States today is the acknowledged leader of the free world; it has global responsibilities and obligations. The United States does not covet territory, nor is it militaristic, but it must have an adequate standing military force to protect its own interests and those of other democratic nations. Our boundaries, so secure for hundreds of years, are now vulnerable to attack from both air and sea. Our frontiers extend to the Far East, to Africa, Europe, Asia, the Caribbean, and the Indian subcontinent. The world has grown much smaller, thanks to technology and human progress. Events that occur in distant countries have repercussions that directly affect the United States economically, militarily, and politically. The dissolution of the Soviet empire in Eastern Europe will affect the US in both the short term and the long term. It is no longer valid to view the world as divided into communist and capitalist camps; other actors play important regional roles, and the balance of power is constantly shifting. Military officers today not only must be proficient in their fighting skills, but they must also have a grounding in international relations and be sensitive to political, economic, and social changes around the world.

The professional officer has an important role in the American way of life. He has dedicated his life to the service of the country. Under extreme circumstances he is willing to place his life on the line for the defense of the nation. His devotion to duty carries a moral as well as a professional obligation. The standards of conduct for the professional officer are well defined and apply both on and off duty. It is often said that serving in the armed forces is a round-the-clock job. The officer is expected to set an example and meet all obligations, both personal and military, in a professional manner. The best interests of the country come before personal considerations. The officer accepts the commission voluntarily and serves at the order of and under the authority of the President of the United States. In taking the oath of office, the officer swears to "support and defend the Constitution of the United States against all enemies, foreign and domestic."

The military profession is dedicated to service, and officers serve with honor. The code of honor must be strictly observed, whether as a cadet or midshipman at a service academy, as an ROTC cadet, or on active duty. The professional officer is expected not to quibble, hedge, or offer excuses, but to carry out his duties to the best of his ability. Once a decision has been made, the officer is expected to support it and carry it out as best he can. An officer's word and signature are his or her bond; he or she will never lie, cheat, or steal, will never disregard or slack off on the military mission for personal gain. An officer keeps promises and meets financial obligations. His or her moral character, integrity, courage, and devotion to duty are above reproach.

During most of our history, the military has been relatively isolated from the civilian community. Military personnel were usually stationed in isolated posts, in foreign countries, or on ships at sea. There was no draft until this century, and not all segments of the population were fully represented in the military. As a result, the public at large knew little about military customs and traditions, and the military in turn knew little about public concerns and perspectives. The situation is quite different today. Close and continuing contacts exist between civilians and military people at innumerable levels. The media—newspapers, TV, radio, and movies—provide extensive coverage of military operations, research and development, budgets, training, educational requirements, and so on. In our democracy the people have a constitutional right to be informed of governmental activities—government in the sunshine. Military undertakings, to be successful, must have the support of the people. Today's professional officer must understand that America's armed forces are indeed a people's armed forces. We learned many lessons from the Vietnam War, but the most important was that no war can be successfully prosecuted without the consent and active support of the public.

The professional officer must accept the responsibilities of American citizenship to the degree that they are compatible with military duties. Since the officer's professional responsibilities require that he or she be nonpartisan, he cannot be publicly committed to one political party or faction as far as official duties are concerned. He or she cannot use an official position as leader, manager, or supervisor to try to influence subordinates. Military officers are the executors of policy; the makers of policy are the duly elected civilian leaders. That does not mean, however, that professional officers should not be interested in or informed about politics. Professional officers may not hold elective office while on active duty, but they are encouraged to participate in the political process. Officers stationed outside the United States are provided with information on both local and national elections so that they can make an informed choice. Voting is an obligation, a responsibility, and a privilege for all American citizens.

The professional officer is encouraged to participate in community activities and voluntary associations both on and off the military installation. Today, more than at any time in the past, large numbers of military officers live in civilian communities. More officers are working in civilian industries, attending civilian universities to obtain advanced degrees, and working in diversified areas away from strictly military duties. This enables them to gain an appreciation of the civilian point of view and helps the civilian community to understand the military point of view. It is a good basis for effective civilian-military relations.

The professional officer today and tomorrow faces unique challenges: how to operate in an environment in which technology advances almost daily and military adversaries have the potential to destroy life as we know it. These challenges are all the more difficult because basically it is against the American character to prepare for future war. We believe that conflicts should be resolved by discussion and reason, with force used only as a last resort. We believe in the basic goodness of humanity and that our form of government and way of life are the best. We believe that government should serve the people, not vice versa. These are noble ideals, but we must remember that they are not shared by all govenments. Our world view works against programs to keep the military fully prepared to respond immediately to military attack. The American tradition and ethic demand that the United States never be an aggressor nation, never attack first, that it conduct warfare in an honorable manner, that it must be threatened significantly before resorting to violence. The public must be kept informed of how the war is being conducted and must be able to see a prompt successful conclusion to the conflict. These imperatives make service in the armed forces challenging and difficult.

Today's professional officers must possess more technical skills than have been required of their predecessors. All the services need officers with engineering, scientific, computer, research, and technical skills. The armed forces and the maritime industries are at the forefront of exploring applications of the newest technologies in space, at sea, and on land. The operation of complicated weapons systems and their supporting units requires officers of a high level of technical knowledge and skill. The tremendous research and development effort of the armed forces also requires highly trained technicians: some pure scientists interested in the development of fundamental disciplines, some with the specific task of understanding scientific developments and relating them to the needs and missions of each service.

To perform its essential function in national defense, each service must have highly qualified, well-educated, dedicated officers capable of executing the multitude of tasks necessary to readiness and to planning and preparing for the future. Probably many cadets who are now attending service academies will one day take on tasks that do not now exist and cannot even be imagined. The services require officers who are intelligent, resourceful, and visionary and who possess the necessary academic and technical credentials.

Important as technical skills and expertise are, however, the main emphasis in all services is on leadership. Being an officer means that your first responsibility is to manage, supervise, and lead the men and women who work for you. Military leadership encompasses two major responsibilities: to accomplish the military mission, and to watch out for the well-being of the people entrusted to your care. No other profession makes demands on its members as does the military in wartime. In extreme circumstances the military officer is called upon to guard the nation and, if necessary, make the ultimate sacrifice to preserve its way of life and its ideals. In time of peace people are often motivated by personal gain and by the needs of survival and security. In time of war, however, the military officer must be able to inspire his or her people to sacrifice self-interest—possibly to sacrifice their lives—to carry out missions for the good of the service and the country. Leadership can be learned, and the service academies start the lifelong process of learning it.

Specific skills of problem-solving, decision-making, planning, goal-setting, communicating, coordinating, supervising, evaluating, motivating, teaching, and counseling are required of officers in all services. Today's professional officer cannot be effective if he or she is not people-oriented. The days of strict authoritarianism and unthinking obedience are long gone, if in fact they ever existed. Today's military members are encouraged to think for themselves and to participate in the decision-making process. The American character does not allow a military whose members are mere automatons in a rigidly structured system such as that of the Soviet Union. American military members contribute their thought and experience for the most efficient operation of their unit.

Today's professional officer must be prepared to operate effectively anywhere along the spectrum of conflict. That could range from limited unconventional war or combating terrorism to all-out nuclear war. The United States policy in the nuclear age is one of deterrence. The armed forces must be strong enough and determined enough to make potential adversaries realize that aggression against the United States or its allies is not worth the risk involved. The likelihood of nuclear war is very low in the foreseeable future; leaders of both the United States and the Soviet Union realize that nuclear war has the potential to destroy life on earth. Still, the possibility exists of an unauthorized or accidental firing of a nuclear weapon, or even of a madman's choice.

The highest probability of conflict in the near future is low-level insurgencies and terrorist action in areas far distant from the United States. Small-scale limited operations such as those in Grenada in 1983 and Panama in 1989 may again occur. Today's officer must be prepared for the possibility of military operations in Europe, Africa, Latin America, the Far East, and the Persian Gulf area. There are strong indications that the military's role in fighting the production and distribution of illicit drugs will be greatly expanded in the next decade. That will be a relatively new mission for the armed forces, and one that will require specialized training. Tomorrow's officer will have to be flexible, be able to respond on short notice, be able to work in any environment, be able to plan quickly and change plans as the situation demands, and be well versed in foreign cultures. It will be a difficult and professionally rewarding challenge.

Much has been written about the character of the United States military. An all-volunteer force is necessarily different in many ways from a force depending on the draft for its manpower. The draft was discontinued in the early 1970s, and arguments persist as to whether it will be necessary to reinstitute it to meet manpower requirements in the 1990s and beyond. Arguments for some sort of a national service requirement for all eligible young people periodically surface. Constantly changing attitudes toward military service and a decreasing manpower pool to draw from in the next ten years may oblige the armed forces to increase recruiting efforts and enlistment incentives to meet goals. It is possible that the service academies will increase efforts to attract quality men and women. It is also possible that public opinion will mount to decrease the size and budget of the armed forces because of the lessening of world tensions, and the armed forces will have more applicants than they need. Whatever the case, it is important for you to know as much as possible about the service academy application procedures and academy life to enhance your chances for acceptance into the academy.

The military constituency is drawn from the society it defends, and society as a whole has certain perceived images about the military. Society in a special sense looks to the military as possessing some of its most precious values. Moral integrity is synonymous with military officership. To give but one example, consider the question of cheating in the universities. Cheating, according to most polls, is generally accepted by the public, and by the age of ten most people have developed a noncondemning attitude toward it. What is the public's reaction to a report of cheating at a university? Generally speaking, very little. What is the public's reaction to reports of cheating at a service academy? The story receives widespread media attention, committees are appointed to investigate the circumstances, immediate changes are called for, and supervisory personnel are usually replaced. Somehow we feel that such things are not supposed to happen at military academies. The public perceives moral integrity as such an integral part of military service that any hint of cheating or dishonesty must be quickly and completely examined and remedial measures taken. Military officers must always exhibit the highest moral standards; there are no exceptions.

Entrusting the security of the country to the military demands that the military demonstrate those values that are precious to society. Public expectations of military officers' performance of duty go far beyond military knowledge and technical proficiency; the public expects military officers to exhibit loyalty, patriotism, obedience, selflessness, and above all, integrity. The profession itself is viewed as noble because it involves protecting the values that citizens cherish most highly. Military officers must be aware of and comply with the public's expectations.

Attendance at and graduation from the service academies provide the dedicated young men and women of this country with training and experience that are invaluable for the rest of their lives. Whether they serve a career in the armed forces, the Coast Guard, or the Merchant Marine, they have become something special. They are some of the country's most valuable assets, and they have also willingly taken on great responsibilities as officers serving the United States.

Chapter **III**

The United States Military Academy

History

Fifty miles from Manhattan, where the Storm King mountain stands guard over the historic stronghold, on the west bank of the Hudson River is one of the world's most famous military installations. When you enter the United States Military Academy, you become part of a tradition almost as old as the United States itself. The first of the service academies, West Point has trained officers for almost 190 years.

West Point's role in the nation's history dates back to the Revolutionary War, when both sides realized the strategic significance of the Hudson River. Had the British gained control of the river, they could have split the colonies in two and defeated both sections. The colonists sought to secure control of the river by occupying the high ground dominating its narrow "S" turn at West Point. Because of its strategic importance, General George Washington considered West Point the most important position on the continent. He had a hand in fortifying West Point in 1778 and transferred his headquarters there in 1779. Continental soldiers built forts, batteries, and redoubts and stretched a 150-ton iron chain across the river to control traffic. Fort West Point was never captured by the British, and it is the oldest continuously occupied military post in the country.

Seeking independence from wartime reliance on foreign engineers and other specialists, leaders of the young nation—

including Washington, Henry Knox, Alexander Hamilton, and John Adams—urged the creation of an institution devoted to the arts and sciences of warfare. In his last letter on December 12, 1799, two days before his death, Washington wrote to Hamilton that the establishment of such an institution "has ever been considered by me as an object of primary importance to this country."

Finally by Act of Congress signed by President Thomas Jefferson, the United States Military Academy was officially established on March 16, 1802, and opened on Independence Day of that year. The initial strength of the academy was five officers and ten cadets. Teachers were few and usually underpaid. There was little discipline. Cadets dressed as they pleased and lived where they chose. There were no prescribed courses of study, and men were graduated after two to six years of service.

The original cadet uniform was blue, similar to the Army officer's uniform of the time except that epaulettes were omitted. A silk hat was worn with the uniform. In 1814 General Winfield Scott was unable to obtain blue fabric and clothed his Army in gray. The Battle of Chippewa in the War of 1812 was won by these gray-clad troops. In honor of that victory, gray was adopted as the uniform of the Corps of Cadets. The cadet full-dress uniform of today is basically the same that has been worn since 1816.

In 1817, Colonel Sylvanus Thayer, the "father of the Military Academy," became superintendent. Thayer, a graduate of the class of 1808, had made a study of the systems of military education in Europe, and he proceeded to put his ideas into effect. During his tenure—1817 to 1833—many characteristic features of the Academy were established: the organization of a Cadet Battalion, the division into classes and sections, the transfers between sections, the weekly reports, the establishment of class rank, the checkbook system, the system of demerits, the amenability of cadets to martial law, the semi-annual examinations, the summer encampments, and the granting of furloughs to members of the second class.

Mindful of the desperate need for engineers, Thayer made civil engineering the heart of the curriculum. He also emphasized small classes, regular study habits, and the requirement that every cadet must pass each course or make up his failure. Colonel Thayer's system was so successful in developing leadership, character, and integrity in the Corps of Cadets that it is still the cornerstone of Academy training. When the coat of arms was adopted in 1898, the Academy motto "Duty, Honor, Country" inscribed thereon was the effort of another generation to put Colonel Thayer's ideal into words.

Through 1865, West Point carried on a dual role as a national military academy and school of engineering. That the graduates were well trained in military sciences is attested to by General Winfield Scott: ". . . but for our graduated cadets, the war between the United States and Mexico might, and probably would, have lasted some four or five years, with, in its first half, more defeats than victories falling to our share . . ." The records of the Civil War reveal that the Confederacy used graduates whenever and wherever possible, including Generals Robert E. Lee and Andrew Jackson. The Union Armies at first had "political" generals, but by the last year of the war all senior commanders were Academy graduates.

By 1866 it was apparent that the Army would need specialized training of officers in particular branches of service to keep abreast of the expanding science of warfare. To meet this requirement, several Army postgraduate schools were established, and West Point came to be regarded as the initial step in the Army's system of education. Accordingly, the curriculum had as its objective general instruction in the elements of each military branch.

In 1902 the Academy underwent a complete structural renovation and a reassessment of the military and academic curricula. Military instruction was changed to practical training in tactics and field exercises. Coordination was developed between instruction and actual field conditions. World War I tested and again proved the soundness of the Academy's curriculum and training.

General Douglas A. MacArthur became Superintendent of the Academy in June 1919. Under his direction the curriculum was updated to take advantage of the lessons of World War I. A combination of general and technical education was provided to give a solid foundation for a professional military career. Increased emphasis was placed on physical fitness.

Intramural sports are an integral part of the Military Academy program.

"Every cadet an athlete" became the goal. In addition to a strenuous program of compulsory gymnastic instruction, every cadet was required to participate in an intramural program of fourteen sports. This recognition of the importance of physical fitness continues today at the Academy, where all cadets are required to participate in sports and the athletic program is guided by the dictum, "Every cadet an athlete, every athlete challenged." Additionally, the administration of the honor system by the cadets themselves, long an unofficial tradition, was formalized with the creation of the Cadet Honor Committee.

During 1939–41 all phases of training were intensified, and again the value of the education and training given at the Academy was abundantly proved by the roles played in World War II by such graduates as Dwight D. Eisenhower, MacArthur, Omar N. Bradley, Henry H. Arnold, Mark W. Clark, George S. Patton, Joseph W. Stilwell, Jonathan M. Wainwright, and many more.

The postwar period again saw sweeping revisions in the West Point curriculum required by the explosive developments in science and technology, the increasing need to understand other cultures, and the rising level of general education in the Army. The Military Academy began to supplement the basic course with elective study programs, allowing the cadets time to follow more specialized interests. In a consistent pattern of change, the Academy has always studied the experiences of war, including the Korean and Vietnam conflicts, to modify and revise its curriculum and teaching methods. The Academy also has incorporated lessons learned from Grenada and Panama into its teaching program.

In 1964 President Lyndon B. Johnson signed a bill increasing the strength of the Corps of Cadets from 2,529 to 4,417. A major expansion of facilities, to keep up with the growth of the Corps, was also undertaken at that time.

One of the most significant developments in recent years has been the diversification of the membership of the Corps of Cadets, resulting in unparalleled increases in the numbers of blacks, Hispanics, Asians, and other minorities. In 1975 legislation was passed permitting women to compete for admission to the service academies.

In 1982 a historic change was made in West Point's academic curriculum with the decision to offer academic majors for the first time. Majors are optional and secondary, in the sense that the broad core curriculum is considered to be every cadet's professional "major." The majors do, however, allow cadets to pursue in greater depth an academic discipline of their choice. Academic and military life at West Point have changed steadily over the years along with the expansion of knowledge and the changing needs of the Army and the nation. Unprecedented advances in technology, communications, weapons, and space and nuclear research have changed the face of war. The Academy is keeping up with these changes, but it is also maintaining its emphasis on leadership, academic excellence, and personal integrity.

Mission of the United States Military Academy

As we have noted, the Academy was founded to train military technicians for all the branches of military service and to encourage the study of military arts and sciences. By the mid-1820s, Colonel Thayer, who held Master of Arts degrees from Dartmouth College and from Harvard University, could feel considerable satisfaction in the progress of the Academy in accomplishing its mission. West Point was beginning to be known as the major scientific school in the United States.

In 1828 Secretary of War Peter B. Porter said, "The Military Academy . . . is scattering the fruits of its sciences . . . not merely to the rest of the Army, but to the youth of our country generally, and the interchange of the theoretic science of this national school with the practical skill and judgment of our civilian engineers, which is now going on throughout the United States, will soon furnish every part of the country with the most accomplished professors in every branch of civil engineering."

West Point and its athletes continued to exert great influence by furthering national growth in the scientific and general education fields until 1870. This was largely because instructional materials prepared by Academy graduates for use at West Point were also used in other schools. Also, many graduates were employed as instructors by civilian colleges. With the tremendous expansion of scientific knowledge during this period, and with the advances in the science of warfare, it became apparent that the Academy could not continue to graduate enough adequately trained Army officers and engineers to meet both military and civilian needs. Gradually, emphasis was shifted to training and education, so that by 1902 West Point was considered the initial step in an Army officer's education; each student would receive general instruction in the elements of each military branch and later pursue postgraduate studies in the various military specialty schools.

The guiding principle of all the service academies, stated in various ways, is that nothing should remain static. West Point and all the other service academies must always provide a military education tailored to meet the changing conditions of a complex world.

As approved by the Department of the Army, the current mission of the Academy is stated as follows: "The mission of the United States Military Academy is to educate and train the Corps of Cadets so that each graduate shall have the attributes essential to professional growth as an officer of the Regular Army, and to inspire each to a lifetime of service to the Nation."

Although the exact wording of the mission changes over time, the following objectives are always inherent in it: *mental*—provide a broad collegiate education in the arts and sciences leading to a bachelor of science degree; *moral*—develop in the cadet a high sense of duty and the attributes of character with emphasis on integrity, discipline, and motivation essential to the profession of arms; *physical*—develop in the cadet those physical attributes essential to a lifetime career as a Regular Army officer; and *military*—provide a broad military education rather than an individual proficiency in the technical duties of junior officers. Such proficiency is, of necessity, a gradual development, the responsibility for which devolves upon the graduates themselves and upon the commands and schools to which they are assigned after being commissioned.

These goals and objectives are, admittedly, a great deal to expect of any young man or woman. However, by careful selection before admittance and by professional training during the four-year curriculum, the Academy does graduate classes each year that meet these high standards. You must be aware of these high standards and high expectations for you at West Point. Tradition is a major component of life at West Point; duty, honor, country will be with you every day.

Facilities

The garrison area of West Point, owned by Stephen Moore, had been occupied by the Army from 1778. General Washington established his headquarters at Moore's house in 1779 and maintained it there that summer. After the war in 1790, the area was purchased by the government for $11,085. When Colonel Thayer became Superintendent, the reservation consisted of 1,795 acres. Additional purchases and the donation of Constitution Island by Olivia Sage in 1909 more than doubled the size of the reservation. From 1938 to 1945 it was further enlarged by the purchase of 11,401 acres, and a gift of 1,040 acres in 1959 increased its size to the present 16,000+ acres. Of this total, about 2,500 acres are in the Post proper. A golf course, a ski slope, and picnic, camping, hunting, and fishing areas are included in the outlying tract. The main cadet facilities—barracks, academic buildings, gym activities center, and chapels—are grouped within easy walking distance. With few exceptions, the buildings are in the Gothic style.

The hub of the cadet area is Washington Hall, the dining hall and headquarters of the Corps of Cadets, surrounded by cadet barracks. Although called "barracks," the cadet dormitories are similar to those at a civilian college. Generally, two or three cadets share a room. Cadets live with other members of their class within their company; a company comprises about 110 cadets from all classes. Women room together within their assigned companies. An extensive building and rebuilding program has been completed to provide facilities to accommodate the Corps and to replace older dormitories. Enforced evening quiet hours help to maintain an atmosphere conducive to study.

Some academic departments, classrooms, and laboratories are located in Washington Hall. Others are located in Thayer, Bartlett, and Mahan Halls near the cadet barracks. A riding hall in earlier days, Thayer Hall today houses a computer center, a television studio, two large auditoriums, and the Military Academy Museum, as well as 98 classrooms. Construction has begun on a phased renovation of Thayer Hall to expand classrooms, laboratories, and faculty offices. The computer center and Military Academy Museum are to be moved to other West Point locations. Nine-story Mahan Hall was completed in 1972. Located within the academic area, the Cadet Library contains 550,000 volumes, reading rooms, seminar rooms, microfilm and audiovisual facilities, and rare book collections.

West Point's modern academic facilities are matched by its athletic facilities. Michie Stadium, home of the Army football team, attracts crowds in excess of 40,000 during picturesque fall football weekends. Adjacent to the Stadium is the newest athletic facility, a multisport complex housing a hockey rink with seating for 2,500 and a basketball arena with a 5,000-seat capacity. The huge gymnasium building contains five gyms, three swimming pools, including an Olympic-sized pool, and numerous other special-purpose rooms for squash, handball, weight training, and combatives. Varsity and intramural athletic events are held in a field house, on indoor rifle and pistol ranges, on a baseball diamond or a rubberized outdoor track. Also within the immediate military reservation are other athletic fields, tennis courts, and outdoor swimming pools.

Southwest of the campus, the reservation's lake-dotted forest highlands provide an extensive military training and recreational area. In the summer, Camp Buckner and Lake Frederick are focal points for field exercises of varying intensity. In addition, Army Reserve Components perform field exercises, Scouts and other civilian groups camp and hike, and local townspeople enjoy the recreational use of Long Pond.

Three separate chapels provide Protestant, Catholic, and Jewish services. The Cadet Chapel houses the world's largest church organ. Eisenhower Hall, the Cadet Activities Center, was completed and opened in 1974. This modern student union contains a 4,500-seat auditorium, a 1,000-seat snack bar and cafeteria, a large ballroom overlooking the Hudson River, a games area, an art gallery, and a spacious foyer for cadets and guests. Grant Hall, Cullum Hall, and the First Class Club provide additional cadet snack and lounge facilities.

If you wish to visit the campus to help you in your decision about applying for the Academy, a Visitors' Information Center is available on new South Post. Along with the government-owned Hotel Thayer, it helps to accommodate the hundreds of thousands of guests who visit West Point each year. West Point is constantly modernizing and upgrading its facilities. It has some of the finest facilities available of any institution in the country.

Entrance Criteria and Application Procedures

Each year the United States Military Academy (USMA) admits approximately 1,400 young men and women. By design, the new cadets come from all sections of the United States and represent nearly every race, religion, and culture in this country. Nurtured by the West Point environment, this diversity of background helps each cadet gain a cultural as well as a rich educational experience. Candidates must meet the requirements specified by law and must be qualified academically, physically, and medically. Each candidate must also obtain a nomination from a member of Congress or from the Department of the Army in one of the service-connected categories described later in this chapter.

Candidates are considered on the basis of academic perform-

This USMA cadet is clearly enjoying her job.

ance (high school record and SAT or ACT scores), demonstrated leadership potential, physical aptitude, and medical qualification. West Point seeks a class composition of top scholars, leaders, athletes, women, and minorities to maintain a diversified collegiate environment and student body. Candidates with outstanding qualifications in one or more areas and those who have social, financial, or ethnic disadvantages that limit athletic, academic, or leadership opportunities receive special consideration for admission.

Profiles of individual classes admitted to USMA reflect how competitive the process of admission has become. The Academy has a highly qualified pool of applicants from which to select. As you can see on page 34, applicants admitted to West Point had impressive high school academic and extracurricular records. Over 85 percent ranked in the first quintile, over 90 percent participated in high school athletics, over 60 percent belonged to national honor societies, almost all participated in high school activities such as student government, debating, dramatics, scouting, and high school newspaper. Scores on the ACT and SAT were well above the national average. You will greatly improve your chances for selection if you maintain an excellent academic record and participate in extracurricular activities, particularly if you fill leadership positions in them. The USMA is looking for well-rounded cadets with leadership potential, and participation in outside activities is a big plus for you in the selection process.

Applying for admission to the USMA, like applying to all the service academies, is a fairly lengthy process. The candidate is well advised to contact the academy early in his or her high school career (sophomore or junior year) and to keep in touch with the officials during the process. All service academies provide catalogs, information bulletins, and application guides to prospective candidates upon written request. Addresses for each academy are given at the end of each chapter.

The following procedure guide is a shortened version of that recommended by the USMA:

PROFILE — CLASS OF 1991

VOLUME OF APPLICANTS

	Men	Women
APPLICANT FILES STARTED	12547	1946
NOMINATED AND EXAMINED	5041	763
QUALIFIED (academically, medically & in physical aptitude)	2565	311
ADMITTED	1204	155

RANK IN HIGH SCHOOL CLASS

FIRST FIFTH	86.3%
SECOND FIFTH	8.3%
THIRD FIFTH	4.4%
FOURTH FIFTH	0.9%
BOTTOM FIFTH	0.1%

AMERICAN COLLEGE TESTING (ACT) ASSESSMENT PROGRAM SCORES*

RANGE	ENG	MATH	NAT SCI	SOC SCI
31–36	2%	39%	47%	13%
26–30	31%	46%	41%	47%
21–25	59%	15%	10%	30%
16–20	8%	0%	1%	7%
11–15	0%	0%	1%	3%
MEAN	24	29	29	26

COLLEGE BOARD ADMISSIONS TESTING PROGRAM (CBATP) SCORES*

	APTITUDE	
RANGE	VERBAL	MATH
700–800	4%	21%
600–699	28%	56%
500–599	52%	22%
400–499	15%	1%
300–399	1%	0%
MEAN	560	640

*Includes only scores used as a basis for admission

ACADEMIC HONORS

CLASS VALEDICTORIANS	99
CLASS SALUTATORIANS	83
NATIONAL MERIT SCHOLARSHIP RECOGNITION	327
NATIONAL HONOR SOCIETY	893

ACTIVITIES

BOYS/GIRLS STATE DELEGATE	353
CLASS PRESIDENT OR STUDENT BODY PRESIDENT	451
SCHOOL PUBLICATION STAFF:	
School Paper Editor or Co-Editor	82
School Paper Staff	205
Yearbook Editor or Co-Editor	92
Yearbook Staff	208
DEBATING	169
DRAMATICS	166
SCOUTING PARTICIPANTS	475
Eagle Scout (men) or Gold Award (women)	104
VARSITY ATHLETICS:	
Participated	1229
Letter Winner	1195
Team Captain	791

GEOGRAPHICAL DISTRIBUTION

The Class of 1991 includes cadets from across the United States, and ten foreign countries, including Bangladesh, Cameroon, Haiti, Honduras, Jordan, Kenya, Malaysia, Malawi, The Philippines and Turkey.

1. Determine whether you meet the requirements and qualifications.
2. Apply for a nomination.
3. Start a file at West Point.
4. Fill out USMA forms.
5. Take the ACT or SAT.
6. Take the Qualifying Medical Examination.
7. Take the Physical Aptitude Examination.
8. Await evaluation and status of application.
9. Prepare for entrance to USMA.

A brief explanation of each step follows.

1. Requirements and qualifications

Requirements

Each candidate must:

- be 17 but not yet 22 years of age by July 1 of the year admitted.
- be a US citizen at time of enrollment (exception: foreign students nominated by agreement between US and another country).
- be unmarried.
- not be pregnant or have a legal obligation to support a child or children.

Academic qualifications

Each candidate should have:

- an above-average high school or college academic record.
- strong performance on the American College Testing (ACT) Assessment Program Exam or the College Board Admissions Testing Program Scholastic Aptitude Exam (SAT).

West Point encourages a strong college preparatory academic background as a prerequisite for admission. Recommended areas of preparation are: four years of English, with emphasis on composition, grammar, literature, speech; four years of math—algebra, plane geometry, intermediate algebra, trigonometry; two years of a foreign language; two years of laboratory science, and a year each of chemistry, physics, and US history. In addition you will find courses in geography, government, and economics very helpful. Some college courses taken prior to entrance may be substituted for similar courses in the Military Academy curriculum.

Medical Qualifications

Candidates must:

- be in good physical health.
- pass a medical exam.

Physical Qualifications

Candidates should have:

- above-average strength, endurance, and agility.
- adequate performance on USMA Physical Aptitude Exam.

2. Apply for a nomination

In the spring of your junior year you should apply for a nomination from one or more of the sources listed below. You must obtain a nomination to compete for admission to the Military Academy. Cadetships are allocated by law to the Vice President; members of Congress; congressional delegates from Washington D.C., the Virgin Islands, and Guam; governors of Puerto Rico and American Samoa; and the Department of the Army. Nominating officials may select up to ten young people to compete for each cadetship vacancy they have. Since some members of Congress will not accept applications for a nomination after a specific date, interested candidates should request a nomination as early as possible. Sample letters are given in the Appendix. Each nominating authority establishes procedures for selection, which may include interviews and questionnaires. The applicant should contact the nominating authority early and keep in touch during the selection procedure. As a minimum, you should apply to your two US senators, your representative in Congress, and the Vice President.

Another category of cadetships is military service–connected nominations. These cadetships are administered by the Department of the Army and are available to sons and daughters of career military personnel of the Army, Navy, Air Force, Marine Corps, and Coast Guard; to sons and daughters of deceased and disabled veterans; to sons and daughters of persons awarded the Medal of Honor; to enlisted members of the Regular Army and the Army Reserve/National Guard; and to cadets enrolled in certain junior or senior Reserve Officers Training Corps (ROTC) programs. A sample letter for this nomination is also included in the Appendix.

3. Start a file at West Point

West Point will open a candidate file for you upon receipt of your completed Precandidate Questionnaire. You may obtain a questionnaire by sending a request to Admissions Office, USMA, West Point, NY 10996-1797. This should be done in your junior year in high school or as soon thereafter as possible. You must have a social security number to establish a file. Your file will be reviewed, and you will be notified if you lack the qualifications to compete for admission.

4. Fill out USMA forms

You will be required to complete and have others complete many administrative forms as you progress through the application. Promptly return all forms you receive from the Military Academy Admissions Office and the Department of Defense Medical Examination Review Board.

5. Take the ACT or the SAT

All candidates must take either the American College Testing (ACT) Assessment Program exam or the College Board Admissions Testing Program Scholastic Aptitude Test (SAT).

Although not required for admission, Advanced Placement Examinations are considered in several subject areas, including mathematics, physics, chemistry, history, and social sciences. Results are evaluated for award of formal credit for course completion or for scheduling into higher-level sections or classes.

ACT

The ACT is given at test centers throughout the world. For information on testing in your locale, consult any high school counselor or write directly to Registration Department, ACT Assessment Program, PO Box 414, Iowa City, IA 52243-0001. To ensure that West Point receives your test results, list the ACT college code number for USMA (2976) on your registration folder. If you are seeking a nomination by a member of Congress also record the special code number 7000 on your registration folder (once for each member of Congress). Each time you use the 7000 code number, you will receive a sealed copy of your ACT score report to be forwarded (or delivered) by you directly to the appropriate member of Congress. It is your responsibility to see that West Point and your congressional contact receive your test results.

SAT

Candidates taking the College Board exam for admission are required to take the Scholastic Aptitude Test. To take the examination, apply to College Board Admissions Testing Programs

(CBATP), Rosedale Road, Princeton, NJ 08541-6200. For additional information, consult your guidance counselor. To ensure that West Point receives your test results, list the CBATP college code number for USMA (2924) on the registration form. To report your scores to your congressional representative, ask your guidance counselor for the representative's CBATP code number and place it on the registration form. If the counselor does not have the code number, contact the congressional representative or the College Board at (609) 921-9000 before completing the registration form. Following these procedures will ensure that your scores are forwarded to West Point and your congressional representative.

6. Take the Qualifying Medical Examination

All candidates must take a Qualifying Medical Examination between June 1 and July 1 of the year immediately preceding the year they plan to enter. For example, if a candidate wished to enroll in July 1992, he or she would have to take the exam between June 1, 1991, and July 1, 1992. One Qualifying Medical Examination meets the application requirements of all service academies and all ROTC scholarship nominations a candidate may receive. The Department of Defense Medical Examination Review Board will schedule your exam and evaluate the results after you have started an admission file. You will receive instructions for taking your medical exam directly from the Review Board.

A limited number of medical waivers are considered and granted each year for candidates who possess outstanding overall qualifications. All inquiries about medical qualification should be addressed to Director, Department of Defense Medical Examination Review Board, Box 3000, US Academy, Colorado Springs, CO 80840-7500.

7. Take the Physical Aptitude Examination

Your strength, endurance, and agility are measured by the Physical Aptitude Examination (PAE). The four events involved in the exam are pull-ups for men and flexed arm hang for women, standing long jump, basketball throw, and shuttle run. You will receive scheduling instructions for the exam from USMA Admissions. You can best prepare for this physical examination by engaging in physical activities such as running, general conditioning exercises, basketball, swimming, strength exercises, and many competitive games. You should also practice specific test items. Your performance will improve greatly as you practice the specific events. The qualifying exam results can be used for any year you apply to West Point. The Army ROTC Physical Aptitude Examination is the only substitute for West Point's exam. If you take the ROTC exam, have the

WEST POINT PHYSICAL APTITUDE EXAMINATION

Candidate Population for a Recent Class

	Pullups	Flexed Arm Hang	Standing Long Jump		Basketball Throw		300 Yard Shuttle		
	(Men)	*(Women)*	*(Men)*	*(Women)*	*(Men)*	*(Women)*	*(Men)*	*(Women)*	*Percentile*
Top	19	59 sec	8'8."	7'11."	90'	62'	55 sec	60 sec	100%
Quintile	15	51 sec							
	13	45 sec			80'	54'			
	12	40 sec	8'0."	6'11."	75'	50'	56 sec	63 sec	80%
	11	36 sec			70'	47'	57 sec		
	9	34 sec	7'8."	6'8."					
		30 sec	7'6."	6'7."	67'	46'	58 sec	66 sec	60%
Middle	8				65'	44'			
Quintile	7	28 sec	7'4."	6'4."	61'	42'	59 sec	67 sec	
	6	26 sec	7'2."	6'2."			61 sec	68 sec	40%
					60'	39'	62 sec	69 sec	
	5	22 sec	7'0."	6'0."					
	4	18 sec			55'	36'	64 sec	72 sec	20%
	3	16 sec	6'8."	5'9."					
Bottom					50'	33'			
Quintile			6'4."	5'5."			65 sec	75 sec	
	1	11 sec	5'8."	4'10."	40'	25'	68 sec	78 sec	

*run on a 25-yard course

results sent to the Director of Admissions at West Point. The accompanying chart details the exercises and the performance of a recent class.

It is hard to overemphasize the importance of physical conditioning at all the service academies. Physical training and exercises are an integral part of each academy's curriculum. If you are not in excellent physical condition when you arrive at West Point, you will make a stressful situation even more difficult. Do not go to West Point thinking you will be able to get into good physical shape there; it is just too difficult with all the other obligations you will have.

8. Await evaluation and status of application

Notification of acceptance is possible as early as November for fully qualified, outstanding candidates who have completed all admissions requirements and received a nomination. Final admission decisions are made in April for candidates entering in July from the available data on each candidate. Files not completed by March 21 are normally closed to further consideration. It is possible that a few candidates will not be notified of acceptance until shortly before entrance in July. Offers of admission are conditional from the time of offer to the date of admission.

9. Prepare for entrance to USMA

Candidates should prepare for the academic, physical, and leadership demands that a cadet faces immediately at the Academy. It is true that you have to "hit the ground running" at the USMA. However, if you have successfully completed the long selection process you are likely to be ready for the challenges of West Point. Continuing vigorous physical conditioning exercises every day prior to reporting in late June is highly recommended. Participation in school and community activities will help you prepare for leadership positions at the Academy.

Two other areas of interest in regard to admissions are the Early Action Plan and the assistance provided by USMA field representatives. The Early Action Admissions Plan is for well-qualified high school seniors who consider West Point their first choice among colleges. Under the plan, applicants can be notified of admission status by January of the entrance year. To be considered you must notify West Point in writing by November 1 and have a complete file ready for evaluation by December 1. Applicants need not have a nomination or the results of the medical examination to apply, but must have both prior to July 1. The Admissions Office will supply additional information on this plan if requested in writing.

The USMA Admissions Office has the assistance nationwide of representatives who provide service to candidates. They include graduates of West Point both in and out of the active Army and US Army Reserve Liaison Officers who have been trained at West Point for this specific program. These volunteers are available to assist you in admissions processing and to answer questions about West Point programs. Write to Admissions to find out who your area field representative is.

There is no better way to learn about the USMA than to visit there prior to your selection. During the academic year, visits (including talks with cadets) can be arranged through the Admissions Office for students who are at least high school sophomores. You should arrange the visit at least two weeks prior to the planned arrival; phone (914) 938-4041. Tours start from the Admissions Office (first floor, building 606) at 10 a.m. Monday through Friday, September through April. In addition, USMA Admissions sponsors cadet visits to many areas throughout the country during Christmas and spring vacations. For additional information on cadet visits, write the Admissions Office.

Academic and Military Programs

The Military Academy, as the only college specifically charged with preparing young men and women for service as officers in the US Army, has a singular educational philosophy. Graduates must be enlightened military leaders of strong moral courage whose minds are creative, critical, and resourceful. The total academic and military training curriculum helps develop those qualities.

Academic Program

The academic curriculum offers a balanced education in engineering and the arts and sciences, while also permitting cadets to pursue academic specialization in a field of study or optional major of choice. The two complements of the curriculum are a broad general core program that is prescribed and an elective program that is individually selected.

The core curriculum is the foundation of the academic program. Equally balanced between the arts and sciences, it provides a foundation in mathematics, basic sciences, engineering sciences, humanities, behavioral sciences, and social sciences. The 32-course core curriculum represents the essential broad base of knowledge necessary for success as a commissioned officer and also supports the subsequent choice of an area of academic specialization. It is, in effect, the "professional major" for every cadet, since it prepares each graduate for a career as a commissioned officer in the Army.

The curriculum complements the core program by providing the opportunity for study in depth through the elective program. Over 28 fields of study and 17 optional academic majors are available. These fields and majors cover virtually all the liberal arts, sciences, and engineering disciplines usually found in a high-quality, selective college or university of comparable size. Cadets may enter any field of study or major without restriction. No special grade point averages are established for entry, and there are no special quotas for particular disciplinary fields.

For most cadets the path to graduation will be the field of study. Pursuit of the field of study requires the student to devote ten of the twelve electives available to courses defined

The Academic Program

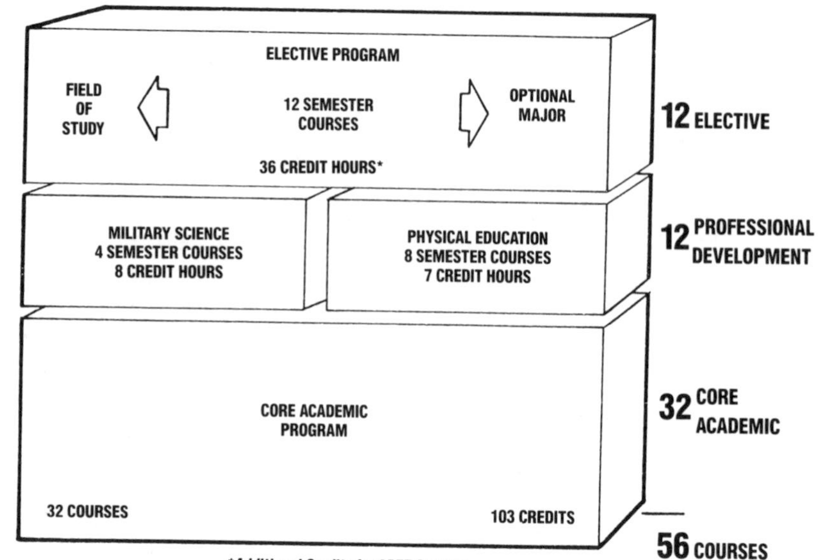

***Additional Credits for ABET Courses**

Mathematics-Science-Engineering Area
Fields of Study/Majors
(All majors are capitalized)

Applied Sciences & Engineering
CIVIL ENGINEERING
ELECTRICAL ENGINEERING
MECHANICAL ENGINEERING
Applied Sciences & Engineering
  Interdisciplinary
Basic Sciences Interdisciplinary
CHEMISTRY
COMPUTER SCIENCE
MANAGEMENT AND ENGINEERING
  MANAGEMENT
MATHEMATICAL SCIENCE
Nuclear Engineering
Operations Research
Physics and ENGINEERING PHYSICS
SYSTEMS ENGINEERING

Humanities and Public Affairs Area
Fields of Study/Majors
(All majors are capitalized)

American Studies
BEHAVIORAL SCIENCES
ECONOMICS
Foreign Area Studies
FOREIGN LANGUAGES (1)
  (One or two; choice of seven)
GEOGRAPHY
Humanities Interdisciplinary
International Affairs
LITERATURE
MANAGEMENT
HISTORY (2)
  (Military, Modern History Field of Study)
Military Studies
National Security & Public Affairs
  Interdisciplinary
Philosophy
POLITICAL SCIENCE

NOTES: (1) Foreign Languages available: Arabic, Chinese, French, German, Portuguese, Spanish, and Russian.
(2) Choice of world or American history.

by the disciplinary field. The remaining two electives are free and can be taken from among the more than 300 electives offered. For cadets who desire to pursue a discipline in greater depth, a voluntary majors program is available. Cadets electing to major must devote all twelve electives to courses prescribed by the disciplinary field, follow a more structured elective sequence, and complete a senior thesis or design project.

The accompanying chart illustrates the main features of the curriculum. To graduate, cadets must successfully complete the requirements of a field of study or major; pass at least 44 academic courses, including successful validation of each course in the core curriculum, eight physical education and four military science courses; and achieve a cumulative grade point average of at least 2.0.

If a cadet is an exceptional student, he or she may enroll in an honors course or advanced individual study in any of the disciplines taught at the Academy. These programs emphasize independent or tutorial work and are excellent preparation for graduate study.

The growing complexity of technology, international diplomacy, and world commitments of the Army has increasingly come to demand that Army officers attend civilian graduate institutions. About 90 percent of Academy graduates who remain in the military more than the required five years attend graduate school through the Army Civil Schooling Program or on a scholarship or fellowship. This is a remarkable statistic.

West Point has some of the finest educational facilities available in the country. Small classes—usually twelve to sixteen cadets—assure individual participation and individual attention. Each cadet is issued a high-performance personal computer for use throughout his years as a cadet and after graduation. The personal computers are linked to the existing academic computer network, providing cadets access from their room to all computing facilities at the Academy, including the central computers in Thayer Hall and the on-line card catalog in the library computer system.

West Point also features a number of special programs and

Comrades in arms in field exercises at the United States Military Academy.

courses for both slow and fast students. Cadets receive extensive counseling throughout the four-year program and particularly during decision time on track choices, elective choices, schedules, course changes, overloads, and so on. Additional instruction is available to students at all times, whether failing a particular subject or not. The individual attention accorded each cadet by the USMA faculty is one of the strongest selling points for attendance at West Point.

Military Program

Military instruction and training are an integral part of the curriculum at West Point. Potential officer-leaders must master fundamental military concepts and skills and know and understand the role and employment of elements of the Army. They must understand and commit themselves to the demanding code of ethics of the American professional soldier.

Cadets receive instruction in the fundamentals of small-unit tactics and leadership through the study of military science and leadership. Physical education and an extensive intramural program prepare cadets for the physical demands of the service and the combat environment. Three summers of field training give cadets repeated opportunities for the practical application of principles learned, while sustaining the high level of fitness demanded of the Army officer.

Classroom instruction and practical experience in the field are combined to develop in cadets the leadership expertise critical to the profession of arms. A strong sense of duty and responsibility is especially valued. In addition to self-discipline, cadets learn to exercise good judgment even when thinking and reacting under mental and physical stress and the demands of time.

Fourth Class Year. During their first day at West Point, the men and women of the incoming class make a rapid transition from civilians to cadets. It is a shock to most of the incoming cadets, and you should be aware that you are going to face a physically, mentally, and emotionally demanding month and a half. The best attitude to have about this orientation is the conviction that no matter how tough it gets, you can make it. Beginning with a particularly challenging first day, you will start the rigorous six-week Cadet Basic Training Program designed to teach you to become both a soldier and a cadet. You will learn to wear the uniform, to prepare your room for exacting inspections, and to participate in parades. Many long hours of physical exercise will prepare you for long marches, land navigation exercises, rifle marksmanship, and tactical maneuvers that are part of the field training in the basic skills of a soldier. You will learn to respond quickly and accurately to orders. This training is to teach discipline, personal pride, confidence, and a sense of duty. It will also provide you with a better understanding of and perspective on the life of an Army recruit.

At the end of this initial training period, in mid-August, the cadets are formally accepted into the Corps. As members of the US Corps of Cadets they have a well-deserved sense of confidence and pride that comes with the knowledge that they have successfully completed probably the most demanding period of their lives. Military instruction during the rest of the Fourth Class year focuses on military heritage, standards of professional behavior, small-unit tactics, and map reading. Playing an important part in every cadet's life is, of course, physical education. May signals the end of the plebe year and brings recognition as upper-class cadets.

Third Class Year. After June vacation, Third Class cadets report to Camp Buckner for eight weeks of military field training. Infantry operations, artillery firing, weapons training, Army aviation, military engineering, field communications, and survival make up most of the training experience. One week is spent at Fort Knox, Kentucky, for familiarization with tank, cavalry, mechanized infantry, self-propelled field artillery, and air defense operations. Emphasis is on small-unit ground combat operations that allow Third Class cadets to apply principles already learned in classroom instruction. The field training is designed to be physically and mentally demanding.

Third Class Military Science courses include instruction in combined arms operations, emphasizing company/team and task force operations, and terrain analysis instruction, emphasizing the ability to gain information on any region in the world using satellite imagery or photographic interpretation.

Second Class Year. In the Second Class summer cadets receive leadership experience in actual Army units, serve as squad leaders at Cadet Basic Training and Cadet Field Training, and participate in military specialty training. Half the class participates in Drill Cadet Leader Training and trains the Army's new recruits at one of the eight US training centers. The remaining half of the class participates in Second Class Detail as squad leaders for the training of the Third Class cadets during Cadet Basic Training.

In both programs, cadets practice many of the skills learned at the Academy and gain invaluable appreciation of the challenges faced by soldiers and leaders of the active Army. At the conclusion of Leader Training, selected cadets participate in various military specialty training courses. Training options include Jungle Training in Panama, Arctic Operations Training in Alaska, Airborne Training at Fort Benning, Georgia, Helicopter Assault Training at Fort Campbell, Kentucky, and Survival, Evasion, Resistance, and Escape Training in Colorado. Many cadets believe these are some of the most valuable experiences of their years at the Academy. The Military Science courses in this year cover the duties of a platoon leader and public speaking.

First Class Year. With the long-awaited First Class year come more privileges and latitude, and much greater responsibility. During the summer before the final academic year, half of the First Class leads the training of the Third Class cadets at Camp Buckner and the new cadets during Cadet Basic Training. The remaining half of the First Class receives leadership experience in actual Army units in Cadet Troop Leader Training (CTLT) and join Regular Army units in Germany, Panama, Alaska, Hawaii, Korea, or regions of the US. With the start of academics, First Class cadets are selected to fill leadership positions from commander of the 4,400-member Corps to leaders of forty member platoons, and staff positions that involve management of all the activities of the Corps of Cadets. The opportunities for planning, organizing, and leading are almost limitless.

Final preparation for the First Class graduation into the Regular Army includes courses in athletic coaching, exposure to practical concerns of military life, and military leadership. The academic instruction and summer training programs are designed to complement and supplement each other. West Point is determined to make the transition from civilian to active duty officer as smooth and productive as possible for both the cadet and the Army.

Cadet Life

Life for the students at service academies is different in many ways from that at civilian colleges and universities. The two major differences are that cadet life is more organized and disciplined and is firmly built around an honor code. It is hard to overemphasize the importance of the Honor Code to the cadet's daily life. The Code states that a cadet will not lie, cheat, or steal, nor tolerate those who do. It is a vital and valued tradition that is the cornerstone of the development of integrity, self-discipline, and motivation essential to an honorable person. The Honor Code is enforced by the cadets themselves, and those who violate it are normally separated from the Academy. There are practical advantages to living under a system in which mutual trust is a reality. When a person's word is a bond and a signature means what it is supposed to mean, life is smoother, simpler, and produces less anxiety.

If one had to describe a cadet's life at West Point in only one word, the word would have to be busy. A typical daily schedule during the academic year is as follows:

6:30–7:10	Breakfast
7:15–11:35	Class or study
11:45–12:30	Lunch
12:30–1:25	Commandant's Time
1:35–3:40	Class or study
3:40–6:00	Intramural, club, or intercollegiate athletics; parades; extracurricular activities; or free time.
6:30–7:05	Dinner
7:05–7:30	Cadet Duties
7:30–8:30	Study conditions/extracurricular activities
8:30–11:30	Study time only
11:30	Taps
12:00	Lights out

The primary emphasis is on academic achievement. Because of the small class size, cadets almost always participate in classes. Outside of the classroom, a multitude of activities are available. The cadets are encouraged to participate in as many date stood a trial period of one year ashore and six months at sea. If he displayed the required officer qualities during this period, he became a midshipman and served at sea for three more years. After another year ashore, he was commissioned lieutenant, US Navy. English, mathematics, geography, French, and Spanish were included in the curriculum. Lectures were given in chemistry, physics, and ordnance. The practice of lismissing and turning back those found deficient in academics vas instituted in 1846, following the June final examinations.

On July 1, 1850, the Naval School became the United States Iaval Academy, and the course of study was condensed to four cademic years. Summer training cruises gave the midshipmen eagoing experience to augment classroom work. Thus, the rerunner of today's basic four-year curriculum and summer uise program first appeared at the Naval Academy over 140 ars ago.

The year 1861 was a sad one for the Academy when mid- pmen of the North and South fell in at their last parade ether. While loyalty to the Union was requested, it was not orced. Schoolmates and classmates parted in deep sorrow, ny to meet later at sea during the bitter struggle of the Civil r. Those who chose to remain loyal to the Union embarked he USS *Constitution* on April 24, 1861, and sailed to New k and then to Newport, Rhode Island, where instruction continued at Fort Adams. The Fourth Classmen (plebes) quartered on the ship, while Third Classmen (youngsters) berthed ashore. The two upper classes were commis- ed at the outbreak of hostilities and assigned to the Union In the meantime, Fort Severn was again utilized as an y post.

der Vice Admiral David D. Porter, the Naval Academy eopened on September 11, 1865. The USS *Constitution*, returned the midshipmen to Annapolis, remained as the n ship until 1870. Admiral Porter supervised the repairs Academy necessitated by Civil War damage and also ed funds for many new buildings. In 1883 the Naval my began commissioning Marine Corps officers.

activities as possible as long as their academic program does not suffer. Cadets may follow interests first stimulated in the classroom such as language, computer, and geology clubs or such activities as chess, rugby, soccer, Cadet Glee Club, radio and TV, and cadet magazines. Many cadets step out of the West Point community to help in veterans' hospitals, juvenile correctional institutions, and young people's organizations. The Cadet Scoutmasters' Council works with local Boy Scout units and annually hosts a camporee. The Cadet Public Relations Council sponsors cadet appearances at junior and senior high schools, for military and civic organizations, and on radio and TV throughout the nation. All cadets compete in some form of athletics on the collegiate or intramural level.

Student government at West Point is the cadet chain of command under the auspices of the Corps of Cadets. Cadets manage the customs and tradition programs (with authority to discipline), the underclass military drill and training programs, and the Cadet Honor System. Cadets also manage most of the intramural athletic programs and other extracurricular functions. The student body is organized as a brigade and is divided into four regiments having 12 battalions and 36 companies; a cadet serves as Brigade Commander. The company is the hub of cadet life. Competition in athletics, parades, and military training is between companies. Long-lasting friendships, social interaction, and tutorial assistance are fostered within the company organization.

Academic, military, financial, and other types of personal counseling are available to cadets at all times. Apart from this professional counseling, cadets can always seek advice from their peers in the cadet chain of command. Legal assistance is available; advice on wills, contracts, taxes, and certain other private legal matters is provided at no cost. Cadets receive complete medical and dental care. Frequent examinations ensure continued excellent health. If hospitalization becomes necessary, cadets receive treatment in the well-equipped Keller Army Hospital at West Point.

The number of vacations (leaves) and the amount of free

time a cadet has depend on seniority as well as military performance. A senior receives about twice as many weekend leaves per semester as a junior. A plebe (freshman) leaves the Military Academy on only four weekends in addition to the Christmas holidays and authorized athletic, extracurricular activity, or cultural trips. Additional weekend passes may be awarded to cadets based upon individual or unit achievement. All cadets take Christmas and summer leaves. Upperclass cadets also take a spring leave. Progressively greater amounts of free time allow cadets to be part of the collegiate subculture at the Academy and to mix with students from nearby campuses.

In return for the cadet's five-year active duty obligation and three-year reserve obligation after graduation and commissioning, the government provides all tuition and room and board costs as well as any other educational fees. A cadet receives more than $6,500 per year. The cadet must pay for a personal computer, his uniforms, and textbooks from this amount. As part of the Regular Army, a cadet is entitled to salary and Army benefits.

If you are thinking about entering West Point, you must be prepared for total involvement in the program. You cannot get through the four years successfully without a lasting commitment to do your best. You must retain a positive attitude about what you are accomplishing and set your sights not only on getting through a day at a time but also on what you will achieve upon graduation. Obtain as much information about the Academy as you can before making your decision. Write for information to Director of Admissions, US Military Academy, West Point, NY 10996-1797. Attending West Point is a valuable experience that will serve you well the rest of your life.

Chapter IV

The United States Naval Academy

History

At the time of the founding of the United
Academy, Navy leaders were trying to interes
idea of establishing a similar school for the
posals were not favorably received at the t
short-lived school was set up in 1803 at the
Yard, it was not until 1838 that a one-year
delphia Naval School, was established at
Philadelphia. This school was essentially for
of examinations for promotion, and atten
was entirely voluntary.

In 1845, George Bancroft, Secretary
President Polk and a strong believer in th
future naval officers ashore, established th
old Army post, Fort Severn, at Annapo
the Severn River empties into the Chesap
was officially founded October 10, 1845,
following year that appropriations were
for "repairs, improvements and instru
Annapolis."

The first superintendent of the Nav
Franklin Buchanan, instilled in the sch
standards necessary to carry it thro
years. During those early years, many
tions, customs, and traditions were f

Undoubtedly the most unceremonious graduation exercise in the history of the Academy befell the class of 1898. While lunching in the mess hall one day in April, the First Classmen were handed their diplomas and ordered to the fleet for duty in the Spanish-American War. When the United States entered World War I, daily recitations were increased and courses shortened to enable the class of 1917 to graduate three months early. The class of 1918 was graduated in June 1917. The three-year course remained in effect throughout the war; not until the class of 1921 was split into 1921-A and 1921-B by class standing did the Academy revert to its four-year course. With the declaration of war following the Japanese attack on Pearl Harbor in December 1941, the Academy again adopted the three-year system. The class of 1942 was graduated in December 1941 and that of 1943 in June 1942. The wisdom of having a source of trained professional officers was amply demonstrated by the war years. Many Academy graduates proved themselves during the war. Many were lost, and they too, contributed to the rich heritage of the Academy.

Following accreditation of the Naval Academy by the Association of American Universities, Congress passed a law in 1933 authorizing the Academy to award the bachelor of science degree to graduates. The Middle States Association of Colleges and Secondary Schools first accredited the Academy in 1947. In 1958 College Entrance Examination Board tests replaced Academy-prepared entrance examinations. Since 1970 candidates have had the option of taking the CEEB tests or the American College Testing Program for admission.

Electives, validation, and "overloads" introduced in 1959 marked the end of the Academy's traditional fixed curriculum. This was followed by the introduction of the Trident Scholars program in 1963; the advent of the Academy's first civilian Academic Dean and the introduction of minors and (for some) majors, both in 1964; and in 1970 the adoption of a required majors program for all midshipmen. Designated bachelor of science degrees in certain engineering majors were first awarded in 1969. Seven engineering majors now lead to designated degrees. All are accredited by the Engineering Accreditation Commission of the Accreditation Board for Engineering and Technology. (EAC/ABET).

An Academic Advisory Board of distinguished Americans, formed by the Secretary of the Navy to advise and counsel the Superintendent on academic affairs, has met periodically at the Academy each year since 1966. Legislation authorizing the admission of women to the service academies "consistent with the needs of the services" was signed by President Gerald R. Ford in October 1975. The Naval Academy admitted its first women midshipmen (81) on July 6, 1976. Now, about 125 women are admitted each year.

Mission of the United States Naval Academy

From 1845 to 1990, as the nation's responsibilities and needs for seapower have grown, the Navy has increased greatly in size and complexity. Keeping pace, the Naval Academy has responded to every challenge, moving from sail to steam and into the nuclear age. The Naval Academy's mission is to prepare midshipmen for duty not only in the active Navy, but also in the active Marine Corps. The missions of the Navy and Marine Corps strongly influence the mission of the Naval Academy as well as the content of the curriculum. The mission of the Navy is to conduct prompt and sustained combat operations at sea in support of national policy. The principal functions of the mission are sea control, power projection, and strategic sealift. The Marine Corps is one of four separate services within the Department of Defense. The Marine Corps and the Navy are separate services within the Department of the Navy. The mission of the Marine Corps is to provide Fleet Marine Forces of combined arms, together with supporting air components, for service with the fleet in the seizure or defense of advance naval bases and for the conduct of such land operations as may be essential to the prosecution of a naval campaign; to develop tactics, techniques, and equipment used by landing forces; and to perform such other duties as the President may direct.

The mission of the Naval Academy is to develop midshipmen morally, mentally, and physically and to imbue them with the highest ideals of duty, honor, and loyalty in order to provide graduates who are dedicated to a career of naval service and have potential for future development in mind and character to assume the highest responsibilities of command, citizenship, and government. At the Naval Academy, just as at the other service academies, the midshipmen have a high degree of patriotism, a sincere desire to serve their country, and a strong sense of purpose.

The Naval Academy does much more than offer a sound college education. The program includes military training, physical education, and the study of the principles of the naval profession. The purpose is to produce self-confident leaders who accept and carry out their responsibilities to the nation they serve and to the men and women entrusted to their command. The ultimate objective of the training that begins at Annapolis is to produce officers who can rise to command—professional officers who are physically strong, mentally aggressive, and who have a solid foundation in science, technology, and the humanities; officers who have high standards of honor, duty, and responsibility.

The English sea captain and author Joseph Conrad described the challenge of command at sea, a role known by many Naval Academy graduates:

> "In each ship there is one man who, in the hour of emergency or peril at sea, can turn to no other man. There is one who alone is ultimately responsible for the safe navigation, engineering, performance, accurate gunfire, and morale of his ship. He is the commanding officer. He is the ship."

Facilities

The Naval Academy is situated on the banks of the Severn River at Annapolis, the historic capital of Maryland. This beautiful setting, where the Severn widens into the Chesapeake

Bay, is ideally suited for training midshipmen afloat and for inspiring them with the seaman's love for salt water. Long recognized as one of the most beautiful historic institutions on the eastern seaboard, the Naval Academy was designated a Historical Landmark in 1963 by the federal government. Each year over one million visitors tour the grounds and buildings of the Academy. Everyone is welcome during daylight hours, and a visitors' center offers guided tours, maps, and information. Annapolis is less than an hour's drive from Washington, DC, and Baltimore, Maryland.

Physical and academic facilities have kept pace with the demands of the curriculum and the Navy's worldwide missions. Fort Severn's original ten acres have grown to 338 acres. Much of this new acreage has resulted from a series of landfills in the Severn River. For example, a landfill completed in 1959 added 56 acres for the athletic fields and new buildings. Construction of a number of buildings in use today, including the chapel, several academic buildings, and core areas of Bancroft Hall, the midshipmen's dormitory, began in 1899 with a congressional appropriation of ten million dollars. Ernest Flagg was the architect; the style is French Renaissance.

Ensuing years saw the addition of new wings to Bancroft Hall; the construction of Mitscher Hall, containing an all-faiths chapel, a chaplain's center, and an auditorium; the construction of Halsey Field House; and the construction of the nearby Navy–Marine Corps Memorial Stadium. A multimillion-dollar renovation of Bancroft Hall was completed in 1965. Beneath the chapel's towering dome lies the crypt of John Paul Jones. Throughout the yard stand other monuments and mementoes commemorating the deeds of great naval heroes and honoring the Navy's finest traditions.

The Naval Academy is in the final phase of a major campus-wide construction and rehabilitation project. Key structures completed in this plan include the science building, Michelson Hall, and the adjoining mathematics building, Chauvenet Hall, both completed in 1968. The 570,000-volume Nimitz Library was completed in 1973. An adjacent engineering building and laboratory complex, Rickover Hall, was completed in 1975. New interiors, including modern classrooms and laboratories, were constructed in Maury, Sampson, and Luce Halls. A full range of facilities and services for student and faculty research, computer-aided education, and educational television are available throughout the academic complex. All new midshipmen are required to purchase a personal desk-top computer during Plebe Summer, which is paid for through deductions from pay. All academic areas are air-conditioned.

The scientific center of the Oceanography Department is the Hendrix Oceanographic Laboratory, built on a pier on the Severn River opposite the Crown Sailing Center and dedicated in 1985. Much of the lab's furnishings and equipment represent contributions by Naval Academy alumni. The 3,500-square-foot multifunction laboratory supports the current emphasis on oceanography in the Navy. In recent years the Robert Crown Center—home of the Intercollegiate Sailing Hall of Fame and

The Navy spirit is nowhere more evident than at sporting events.

a new waterfront headquarters for the Academy's sailing program—has been completed.

A new student activity center in Dahlgren Hall including an indoor skating rink, a cafeteria, lounges, and game rooms has been completed. Construction of the new athletic complex, Lejeune Hall, featuring Olympic-style swimming and diving facilities and a six-ring wrestling arena, was completed in 1982. Macdonough Hall, the Academy's oldest multipurpose athletic complex, has been redesigned and rebuilt from the walls in. Navy lacrosse, soccer, gymnastics, fencing, and boxing are to be headquartered there. The center for daily living is Bancroft Hall, one of the largest dormitory complexes in the world. Stretching over many acres, it accommodates the entire 4,600-member brigade. All the basic amenities for daily living, as well as many for recreation, are found in Bancroft Hall. It contains such facilities as tailor shops, barbershops, post office, medical and dental offices, bookstore, recreation rooms, library, and assembly hall. It also contains the largest dining room in the world, seating the entire Brigade at one time.

The Naval Academy Museum serves as an inspiration to midshipmen by providing tangible evidence of some of the most famous and exciting episodes in our nation's history. Its collection of more than 50,000 individual items offers a unique educational experience, providing both faculty and midshipmen with a valuable and convenient reference source for the study of naval history.

Entrance Criteria and Application Procedures

Because many of the entrance criteria and application procedures are prescribed by law, similarities are common in the admission processes at the service academies. Those similarities are helpful if you are applying to more than one service academy. The one exception in application procedures is to the Coast Guard Academy, for which you are not required to obtain a nomination. Candidates are not required to take separate medical or physical aptitude examinations for each service academy; the results of these exams can be used by any service academy. Unless you have definitely decided that you will attend only one service academy, it's a good idea to apply to several. You should then start and maintain a separate folder on each academy; in each folder include all correspondence, information bulletins, data sheets, catalogs, requirements, and important dates pertaining to that institution. It is possible to be offered appointments to several service academies. After studying the programs and opportunities offered at each school, you can make your best informed decision.

Each year, the Naval Academy Admissions Board selects approximately 1,400 candidates for the freshman class. They come from every state in the Union and from all the varied backgrounds of American life. The Naval Academy encourages this diversity and recognizes the value of a Brigade enriched by members of every race, creed, and culture found in the nation. Students from minority groups are strongly encouraged to apply for admission. The number of minority students in recent entering classes has risen sharply: 262 midshipmen in the class of 1993 (about 18 percent of the class) are members of minority groups.

There are four general eligibility requirements for all candidates. Candidates must be of good moral character. They must be at least 17 years of age and not yet 22 years of age on July 1 of the year of admission. They must be unmarried, not pregnant, and have no dependent children. Finally, except for the limited quotas of foreign midshipmen specifically authorized by Congress, candidates must be citizens of the United States.

Prospective candidates who meet these requirements then must obtain a nomination; qualify scholastically, medically, and in physical aptitude; and be selected for entry. Scholastic qualification is determined by the Naval Academy Admissions Board based on the candidate's school transcript and rank in class, SAT or ACT scores, school officials' recommendations, extracurricular activities record, and other evidence of character and leadership potential.

The Naval Academy is heavily oriented toward technical

skills. You should pursue studies in high school that will prepare you for a rigorous college program with emphasis on scientific and engineering subjects. The quality of the work is important. Seniors who stand below the top 40 percent of their high school class have limited chances for admission. In fact, 82 percent of the members of the class of 1993 ranked in the top 20 percent of their high school classes. Although the Naval Academy does not have rigid requirements concerning the subjects that must be included in the school record, a candidate is strongly encouraged to include the following subjects in the high school or prep school curriculum: mathematics, four years, including trigonometry; English, four years; modern foreign language, a minimum of two years; English or world history, one year; chemistry, one year; physics, one year; introductory computer and typing courses. About 100 members of each plebe class finish one or more semesters of college before admission to Annapolis. They may earn advanced placement in their course of study. Applicants must furnish transcripts of any college work they have taken. Again, the quality of the work is extremely important. All students must enter the Academy as plebes and complete four years of education and training.

To demonstrate your ability to meet the physical and time demands of four years at the Naval Academy, you should take part in athletic and nonathletic extracurricular activities. Since every midshipman is involved in physical exercise every day at the Academy, it is important that you get in good physical shape while in high school. Do not wait until Plebe Summer and then go on a crash program to try to get in shape. You will also save yourself a lot of harassment from the upper classmen if you are in good physical condition. The Academy is interested in leadership potential and your ability to manage your time when comparing your application with all other applications. They will examine your nonathletic activities and records of part-time employment or prior military service to evaluate your versatility and ability to accept responsibility. At the

Naval Academy, like all the other academies, use of illegal drugs or abuse of alcohol will almost certainly eliminate you from consideration.

Applicants should submit a Precandidate Questionnaire to the Naval Academy in the spring of their junior year, or as soon thereafter as possible. The Academy opens a preadmission file upon receipt of the questionnaire, and an initial evaluation is shared with the applicant. Information in this file is also used by the Academy to give the applicant's Representative and Senators periodic status reports, which include an evaluation of the applicant, the results of the medical examination, and other information that may assist the applicant in being selected for congressional nomination. The Precandidate Questionnaire should be requested from the Director, Candidate Guidance Office, US Naval Academy, Annapolis, Maryland 21402-5018.

Candidates must take either the College Entrance Examination Board Scholastic Aptitude Test (SAT) or the American College Testing Program (ACT). These tests may be taken anytime they are offered until February of the year of admission to Annapolis. Students interested in applying for admission are strongly urged to take the tests (in addition to the PSAT) in their junior year to help the Naval Academy and members of Congress make an early evaluation of their candidacy. The average of all SAT/ACT test scores taken after December of the junior year is used by the Academy in evaluating a candidate's scholastic qualifications for admission. The college code numbers 5809 and 1742 for the SAT and ACT respectively should be used in forwarding test scores to the Naval Academy. This request is normally made at the time the tests are taken, but can be made later. If the scores are recorded on the high school transcript, they will be accepted as official scores if the transcript is validated by and sent directly from the school. Arrangements to take the SAT or ACT tests can be made through high school guidance counselors or by writing directly to the College Entrance Examination Board,

Box 592, Princeton, NJ 08541, or 1947 Center Street, Berkeley, CA 94704; or to Registration Department, American College Testing Program, PO Box 414, Iowa City IA 52243.

Obtaining a Nomination

All applicants must have a nomination from an official source to be considered as a midshipman by the Naval Academy. Applicants should apply to all sources for which they are eligible. These always include a US Representative, two US Senators, and the Vice President. Although nominations are not due at the Academy before January 31, many members of Congress evaluate candidates during the summer and some nominate in early fall. Therefore, you should apply for a congressional nomination during the spring of your junior year in high school. Applicants for military service–connected nominations should apply directly to the Naval Academy after July 1 of the year prior to admission. All but a few of the candidates are notified of their admission by the end of February of their senior year. Nomination sources include the following:

US Senators, Representatives, Delegate to Congress from the District of Columbia, Resident Commissioner of Puerto Rico. Each may have five midshipmen attending the Academy at any one time. Ten nominations may be made for each vacancy. It is not necessary to know the official personally. Sample letters of application are given in the Appendix.

Presidential appointments are available to 100 midshipmen each year. These competitive appointments are limited by law to children of career officers and enlisted personnel of the armed forces, including the Coast Guard, who have been serving continuously on active duty for eight years and are still serving, or who are retired with pay. Application should be made after July 1 and before February 15 to the Superintendent, US Naval Academy (Attn: Nominations and Appointments Office), Annapolis,

MD 21402-5019. A sample letter is given in the Appendix.

The Vice President may have five midshipmen nominated from the US at large attending the Academy at any one time. This nomination is very competitive. Application should be made no later than November 1 to the Office of the Vice President, Washington, DC 20501. No correspondence (high school transcripts, supporting letters) should be sent to the Vice President's office. Those materials should be sent to the Admissions Office, US Naval Academy (Attn: Vice Presidential Applicant), Annapolis, MD 21402-5017.

Nominations are available in several other categories: the Governor of Puerto Rico, the Administrator of the Panama Canal Zone, and the delegates to Congress from Guam, the Virgin Islands, and American Samoa; members on active duty in the Navy or Marine Corps with over one year of service; members of Naval and Marine Corps Reserve units; members of Naval and Marine Corps senior and junior ROTC units; and honor nominees from honor Naval and military schools. Two other categories are quotas for children of deceased or disabled veterans, prisoners of war, or servicemen missing in action and for children of Medal of Honor winners. Applicants should write to the Superintendent, US Naval Academy (Attn: Nominations and Appointments Office), Annapolis, MD 21402-5019. Sample letters are given in the Appendix.

Each year the Naval Academy selects for admission several hundred qualified congressional-type alternate nominees (including Vice Presidential, Samoa, Guam, etc.) and competitive nominees (Presidential, Regular Navy and Marine Corps, NROTC, etc.) to bring the size of the entering class up to authorized strength. By law, the first 150, plus three fourths of any others so appointed, must be congressional-type nominees. No special application for these additional appointments is necessary, since all qualified nominees are considered automatically by the Admissions Board.

Classes are small at the Naval Academy, giving everyone an opportunity for full participation.

In meeting the "needs of the service" for women officers, the Naval Academy is authorized by the Navy to admit about 125 women each year. Women compete for the same nominations as do men. The number of women who may be appointed from each of the sources of nominations is proportional to the total number of appointments authorized by law for those sources. Women have done very well at the Naval Academy. In 1984, for the first time in the history of the service academies, a female midshipman graduated first overall in the class.

All candidates must pass a very thorough medical examination. Physical requirements for the services are not identical, although all the physical exams are conducted under the auspices of the Department of Defense Medical Examination Review Board (DODMERB). The examination is designed to ensure that candidates have the physical fitness to carry out the rigorous demands of the Naval Academy program. Physically fit candidates in good health with normal vision seldom have difficulty in passing the examination. Candidates having less than 20/20 uncorrected vision should note the special requirements for those who wear eyeglasses or contact lenses. Persons with defective color vision cannot be admitted. The Naval Academy determines the priority of medical scheduling for Academy applicants. This is necessary because of the limited capacity of examining facilities available throughout the nation. For some candidates, this means that they will not be scheduled for a medical examination until they are found to be otherwise fully qualified for admission and in line to be considered for an appointment.

Candidates must also pass a Physical Aptitude Examination (PAE) to qualify for entry. This examination tests coordination, strength, speed, and agility. It consists of four tests: pull-ups (men) or flexed arm hang (women), a standing long jump, a kneeling basketball throw, and a 300-yard shuttle run. This examination is conducted separately and is not part of the medical examination. It may be given by a teacher or school official holding a degree in physical education or by any commissioned officer on active duty. Beyond the test requirement, candidates are strongly advised to be in the best possible physical condition when they enter the Academy in early July. The first summer is very demanding, starting with the first day. Endurance and upper body strength are particularly important. Cross-country runs, weightlifting, isometric exercises, swimming (ability to swim 100 yards and to tread water or float for 20 minutes), push-ups and pull-ups, and (for women) the flexed arm hang are valuable conditioning exercises.

All candidates holding official nominations are notified of their qualification status by April 15. Offers of appointment are made on a continuing basis from mid-October to June. Fully qualified candidates who have not been offered an appointment by May 1 probably will not be selected. All candidates who have been offered an appointment to the Naval Academy with the entering class have an opportunity to visit the Academy in mid-May for a full day of orientation briefings and tours. The new class is admitted in early July.

The following chart gives profile statistics for the class of 1993:

Applicants and Nominees

Applicants (includes nominees)	15,057
Number of applicants with an official nomination	6,351
Nominees qualified scholastically, medically and in physical aptitude	2,364
Offers of admission	1,709
Admitted	1,403

College Board Scores

(Scholastic Aptitude)

Score Range	*Verbal*	*Math*
700-800	3%	34%
600-699	38%	54%
500-599	48%	12%
400-499	11%	0%
Average scores	576	665

American College Testing Program Scores

Score Range	*English*	*Math*
30-36	6%	66%
25-29	52%	28%
20-24	40%	5%
15-19	2%	1%
Average Scores	25	31

Rank in High School Class

First fifth	82%
Second fifth	14%
Third fifth	4%
Fourth fifth	0%

Previous College and Prep School

The Class of 1993 includes 29 percent (400) from college and post-high school preparatory programs; 14 percent (202) from the Naval Academy Preparatory School; 6 percent (85) from private preparatory schools; and 8 percent (113) who have completed at least six months of study at a college or university.

Geographical Distribution

Midshipmen were admitted from every state in the nation. The Class of 1993 also includes 13 international students, including midshipmen from the following countries: Honduras, Jordan, Japan, Lebanon, Peru, Singapore, Thailand; and two each from Jamaica, Malaysia and the Philippines.

School Honors and Activities

	Percent
Student body/council/government president or vice president	10
Class president or vice president	13
School club president or vice president	28
School publication staff member	26
National Honor Society member	58
Varsity athlete	85
Varsity letter winner	79
Dramatics or debating participant	67
Leader of musical group	9
Eagle Scout	9
Boys/Girls State or Nation representative	20
ROTC, NROTC, AFROTC cadet	7
Sea Cadet	1

Minorities and Women

The Class of 1993 includes a total of 262 minority midshipmen with ethnic backgrounds as follows: Black (89), Asian/Pacific Islander (46), Filipino (24), Hispanic (89), and Indian/Native American (14).

One hundred thirty-one women midshipmen were admitted with the Class of 1993.

Sons and Daughters of Alumni

The Class of 1993 includes 63 sons and 8 daughters of Naval Academy alumni (5 percent of the class.)

Note how highly qualified the accepted applicants are. Over 1,100 ranked in the top 20 percent of their class; over 1,150 were varsity athletes. The SAT and ACT scores were especially high in math. You must work hard to ensure that your qualifications are excellent to give yourself the best chance to be selected.

If you are not selected for a direct appointment to the Naval Academy, the Admissions Board automatically considers you for selection to the Naval Academy Preparatory School or recommends you for a Naval Academy Foundation preparatory school scholarship. You do not apply directly for either preparatory program.

Located in Newport, Rhode Island, the Naval Academy Preparatory School (NAPS) offers a ten-month college preparatory course to active duty and reserve Navy and Marine Corps enlisted men and women who are seeking Naval Academy appointments. The program is designed to strengthen the academic background of incoming candidates. Navy and Marine Corps personnel who apply but are not admitted to the Naval Academy are automatically considered for admission to NAPS. The Academy also identifies a number of promising and highy motivated civilian candidates who are not successful on their first attempt at admission and offers them the opportunity to enlist in the Naval Reserve for the express purpose of attending NAPS.

The US Naval Academy Foundation, Inc., also helps promising candidates who are not appointed in their first try for admission. This nonprofit organization awards a limited number of scholarships for post–high school preparatory studies to enhance those candidates' qualifications for admission. The Naval Academy's Admissions Board automatically recommends candidates to the Foundation for sponsorship consideration. No special request is required. Candidates accepted for sponsorship by the Foundation usually have submitted their applications to the Naval Academy before March 1 of their senior year in high school. A limited number of applications will be considered after that time. Cash grants are made to the participating junior college, prep school, or college selected. Parents of applicants are expected to contribute within their means. Each entering class includes about 100 Foundation scholarship candidates. Candidates who successfully complete the Foundation program are normally offered an appointment to the Naval Academy.

You should obtain as much information as possible about all the service academies before making your decision. The Naval Academy has many sources you can tap for information. In Annapolis, the Candidate Guidance Office offers detailed information about the program and admissions procedures. Call the office tollfree at 1-800-638-9156 or write to:

Director, Candidate Guidance
United States Naval Academy
Annapolis, MD 21402-5018

If you live near the West Coast, you can contact a Candidate Guidance branch office in California by calling (415) 966-5931 or writing to:

Tom Teshara
Naval Academy Information Office
Naval Air Station
Moffett Field, CA 94035-5017

The Candidate Guidance Office also coordinates a nationwide network of Naval Reserve officers and civilians who are trained as Naval Academy Information Officers, called Blue and Gold Officers. These officers, located in every state, are well qualified to help you on all aspects of the Naval Academy. You can call the tollfree number at the Candidate Guidance Office to obtain the name of the nearest Blue and Gold Officer.

Academic and Military Programs

The central objective of the Naval Academy, like that of other service academies, is to develop professional officers.

The curriculum blends professional subjects with physical development and a combination of required and elective courses similar in many respects to the programs of other colleges. With the aid of academic and military advisers, midshipmen design their own blend of the curriculum's three elements:

- core requirements in engineering, natural sciences, the humanities, and social sciences, assuring that graduates are able to think, solve problems, and express conclusions clearly;
- core academic courses and practical training teaching the professional and leadership skills required of Navy and Marine Corps officers;
- an academic major in a subject chosen by midshipmen to develop their individual interests and talents.

The Academy offers major programs in eighteen subject areas: eight in engineering; six in science and mathematics, including computer; and four in humanities and social sciences. All graduates are awarded bachelor of science degrees.

Of the majors programs, seven (requiring 148 to 150 hours) lead to designated degrees: Bachelor of Science in Aerospace Engineering, Electrical Engineering, Mechanical Engineering, Marine Engineering, Naval Architecture, Ocean Engineering, and Systems Engineering. These majors are accredited by the Accreditation Board of Engineering and Technology (ABET).

Eleven majors programs (requiring about 144 hours) lead to undesignated Bachelor of Science degrees. Those majors are Chemistry, Mathematics, Computer Science, Oceanography, Physics, Physical Science, Economics, General Engineering, English, History, and Political Science. Honors Programs (148 to 150 hours) in four of these majors—Economics, English, History, and Political Science—lead to designated degrees "with honors."

A midshipman's choice of major influences the number of related and supporting courses required in science, engineering, mathematics, or foreign languages. Majors in the scientific-technical fields require more courses at higher levels in mathematics, science, and engineering, for example, than do nontechnical majors. The scientific-technical programs require no foreign language. On the other hand, programs in social sciences, international studies, and English do require foreign language study, as well as four semesters of mathematics, one year of chemistry, and one year of physics.

You may be sure of your study preferences and your academic aptitude when you enter the academy; however, midshipmen as a rule are not ready to make a firm selection of a major at the beginning of the plebe year. They may have a general idea of the area of interest without being sure what major they should take. They may not yet know whether their talents lie in a technical or a nontechnical field. And, as often happens during plebe year, they may discover that their real interests and abilities do not fit the requirements of the major fields they first considered. For these reasons, selection of a major is delayed until near the end of the Common Plebe Year, when the midshipman has completed nearly two semesters of fundamental courses. Approval by the Academic Dean of each midshipman's choice of major is limited only by the availability of institutional resources. The Naval Academy is, fundamentally, an engineering school, and it is expected that the majority of students will be engineers or technical majors. For those exceptional students with unique backgrounds, overriding interests, and commitments to other fields of study, a broad majors program is offered in those nontechnical sciences and humanities that can be reasonably related to the naval profession.

During the first year at the Naval Academy, midshipmen are placed in a program of study at a level suited to ability and academic background. Plebe-year courses are broad enough in scope to provide a sound basis for selection of a major during the latter part of the year. At the same time they contain an element of commonality that enables midshipmen to progress into any of the majors offered. They include the naval science courses that start professional development. The typical academic schedule for plebes includes five to six academic courses in each of the two semesters. Generally, plebes take Calculus I, Chemistry, Fundamentals of Computing, Leadership, and Rhetoric and Introduction to Literature in the first semester; in the second semester they usually take Calculus II, Chemistry, American Naval Heritage, Fundamentals of Naval Science, Introduction to Naval Engineering, and Rhetoric and Introduction to Literature. After the first year, a cadet's courses are individualized to pursue his or her major.

The major academic areas under the direction of the Academic Dean are organized into four divisions—Engineering and Weapons, Mathematics and Science, US and International Studies, and English and History—each headed by a Navy captain or a Marine colonel. A fifth major academic area, the Division of Professional Development, is under the cognizance of the Commandant of Midshipmen. These divisions are further divided into academic departments, seventeen in all, which serve as focal points for the administration of the majors program and for the continuing review and development of the curriculum. The departments are chaired by civilian or military members of the faculty.

The calendar year is divided into two semesters and a summer term. The academic year consists of two semesters, each of approximately sixteen weeks of instruction and one week of examinations. The academic routine includes five days of classroom, laboratory, and study periods each week. Small classes, averaging twenty midshipmen, provide ample opportunity for active classroom participation by each midshipman and for individual attention.

The military program at the Naval Academy gives midshipmen a broad academic and professional foundation upon which to build competence in any warfare specialty they may elect to pursue after graduation: surface warfare, aviation, the submarine service, or the Marine Corps. Professional development of midshipmen starts on the first day of Plebe Summer and continues through graduation four years later. You should be aware of and prepared for this "culture shock" when you report to Annapolis for Plebe Summer. You must be in excellent physical shape, have an attitude that you are going to succeed, and be able to persevere. It will help to tell yourself that thousands and thousands of other young men and women have made it through the program, and you can too. The military training includes professionally oriented classroom studies (fifteen courses during the four years), drills, and practical training conducted during the summer at shore bases and at sea in units of the Fleet. Among the subjects covered are instruction and training in navigation, seamanship and tactics, naval engineering, naval weapons, leadership, and military law. Professional development is monitored and graded throughout the years at Annapolis. These grades are considered along with grades earned in other areas of the curriculum to determine class standing at graduation. A brief listing of the professional training follows:

Fourth Class Summer—Introduction to Seamanship, Physical Education Orientation and Indoctrination, Small Arms Training and Orientation, Indoctrination and Introduction to the Naval Academy and the Navy, Fundamentals of Navy Hygiene, and Basic Cardiac Life Support training.

Fourth Class Year—Physical Education, Infantry Drill, courses in naval engineering, leadership, and naval science.

Third Class Summer—At-sea training for one month or more. Midshipmen are sent to units of the Fleet on both coasts of the US, as well as the Sixth Fleet in the Mediterranean and the Seventh Fleet in the Pacific. Some midshipmen cruise on the Naval Academy's yard patrol craft to various ports and training facilities along the Atlantic seaboard. Third classmen are introduced to Navy life at sea, to shipboard organization and relationships.

Third Class Year—Physical Education, Infantry Drill, Navigation, Naval Engineering and Naval Science courses, and Leadership.

Second Class Summer—Aviation, Submarine, Surface

Line, and Marine Corps Orientation. Professional training is conducted at bases away from the Academy as well as a four-week course conducted at Annapolis on afloat operations and tactics, naval tactical warfare, laws of armed conflict, and public speaking.

Second Class Year—Physical Education, Infantry Drill, courses in Navigation, Naval Weapons Systems, Naval Engineering, electricity, electronics, leadership, management techniques, problem-solving, and decision-making.

First Class Summer—At-sea training for one month or more. First Class midshipmen actively participate in the duties and responsibilities of a junior officer at sea.

First Class Year—Physical Education, Infantry Drill, Weapons, courses in warfare systems design and military justice and military law.

Professional Competency Review (PCR)

The PCR consists of a series of comprehensive examinations administered to midshipmen of each class during the spring semester. The PCR measures whether midshipmen are making satisfactory progress toward achieving the level of professional competence required for graduation and commissioning. It provides an opportunity for midshipmen to evaluate annually their own professional strengths and weaknesses. The PCR also provides feedback on the effectiveness of the professional curriculum, and it is helpful to the Naval Academy to measure how well goals are being achieved.

The academic programs at the service academies are some of the finest offered in the US. The Naval Academy has a highly respected faculty, excellent facilities, and modern equipment. Individual instruction, counseling, and guidance are always available for the midshipmen. Although decisions regarding academic programs must be their own, midshipmen have ample opportunities for consultation with faculty members. During the first few weeks at the Academy, they receive about twenty hours of group and individual counseling on all aspects of the

Sailing activities are engaged in by midshipmen during all four years at the Academy.

curriculum. They also take a number of achievement tests to help determine the levels at which studies should begin. The Academy has programs to allow each student to achieve whatever level he is capable of attaining. All midshipmen admitted to the Academy have the ability to accomplish all its academic requirements.

Graduation Requirements

To qualify for graduation midshipmen must:

1. Complete the courses specified for the assigned major.
2. Complete a minimum of 140 credit hours, of which a minimum of 18 credit hours, exclusive of the required English courses, are in the humanities and social sciences.
3. Achieve a cumulative quality point rating (CQPR) of at least 2.0, a C average.

4. Meet required military professional standards in professional studies and at-sea training.
5. Meet required standards of military performance, conduct, honor, and physical education.
6. Accept a commission in the US Navy or Marine Corps if proffered.

Midshipman Life

Life at the Naval Academy is well organized, purposeful, highly structured, and busy. During the academic year first priority is given to studies, and each midshipman has ample time for out-of-classroom study and research. On weekdays, following the last class of the day, midshipmen participate in intramural or varsity sports and extracurricular activities. During the summer the emphasis swings to professional training, and upperclassmen engage in a program of cruises at sea or in indoctrination visits and training at selected naval shore stations. Upperclassmen also enjoy an extended summer leave.

Moral development is a vital part of the Naval Academy's program. There is an Honor Concept to support and live by. Its standards are high and unequivocal, and every midshipman is expected to meet them. It is assumed that the midshipman will not lie, cheat, or steal, but much more is demanded. Midshipmen are taught formally and by example to recognize the common good, to build a community that shares concern for its members, and to make difficult ethical choices. They learn to do their duty, to take careful oversight of their subordinates, and to meet high standards of moral leadership on all occasions. Such attitudes are developed and tested throughout the Academy's military, educational, and athletic programs.

To accomplish the uniquely military aspects of the Naval Academy's mission, the student body is organized into the Brigade of Midshipmen. The Commandant of Midshipmen, a rear admiral or a senior Navy captain, commands the Brigade. The departments reporting to the Commandant include the

Division of Professional Development, the Brigade Officers, the Physical Education Department, the Brigade Chaplains, and the Midshipmen Supply Department. The Brigade Officers are six battalion officers of the grade of Navy commander or captain or Marine Corps lieutenant colonel, and 36 company officers who are Navy lieutenants and lieutenant commanders and Marine captains and majors. These officers are in close daily contact with the midshipmen in their living quarters. For purposes of military training and administration, the Brigade of Midshipmen is divided into two regiments, each divided into three battalions. The six battalions are each divided into six companies. Midshipmen of all four classes are assigned to each company—the basic military and organizational unit for numerous competitive activities during the year. Each of the military units, from the Brigade down to the 36 companies and their subordinate platoons, is under the command of a midshipman first class, aided by midshipmen staff and assistants. Midshipmen are selected for these commands and staffs on the basis of leadership qualities and order of class standing.

No matter what your background, chances are the living arrangements at the Academy are different from anything you have experienced. The day begins with reveille and ends with lights out. You stand watches, march to meals, and wear uniforms for almost everything you do. You and your roommates must keep your room ready for military inspection at any time and keep your uniforms in regulation condition. Demerits are awarded for a room or uniform that is not in proper order or squared away. All midshipmen live in Bancroft Hall, a huge dormitory complex. You are assigned to a room with one or more other midshipmen and live in close proximity to about 110 other midshipmen in your company. Each company has its own living area, called the wardroom, for meetings and recreation.

Plebe year, starting with the first day of Plebe Summer, is a demanding period requiring midshipmen to stand on their own two feet, to produce under pressure, to respond promptly and intelligently to orders, and to measure up to the highest standards of conduct and morality. The Plebe Summer is designed to turn civilians into midshipmen. It is an abrupt entrance into the military way of life. Soon after entering the gate on Induction Day, you are put into uniform and taught how to salute by the First Class midshipmen ward officers who lead the plebe indoctrination program. For the next seven weeks you start your days at dawn with an hour of rigorous exercise and end the days long after sunset. You have to manage your time and your emotions. Again, it is extremely important that you be in good shape physically before you report for Plebe Summer. If you are the least bit overweight or out of condition, you will make a demanding period even more difficult. When you have successfully completed Plebe Summer and begin the academic year in late August, your routine for the next four years will look like this on weekdays:

6:30	Arise for personal fitness workout and breakfast
7:00	Reveille (all hands out of bed)
7:15–7:30	Special instruction period for plebes
7:35	Personnel formation for muster and inspection
7:55–8:45	First period
8:55–9:45	Second period
9:55–10:45	Third period
10:55–11:45	Fourth period
12:10	Noon meal formation
12:20	Noon meal for all midshipmen
1:15–2:05	Fifth period
2:15–3:05	Sixth period
3:30–6:00	Varsity and intramural athletics, extracurricular and personal activities; drill and parades twice weekly in the fall and spring
6:30	Evening meal formation
6:40	Supper
7:00–8:00	Extracurricular and personal activities; lectures

8:00-midnight	Study period for all midshipmen
11:00	Lights out for plebes

Midshipmen are currently paid over $500 per month, commencing on the date of admission. This salary provides funds for uniforms, books, equipment, laundry, and income tax, as well as personal needs. By graduation, midshipmen will have accrued savings averaging $1,500. Typically, this is used to purchase additional uniforms and to help them get settled at their new station. Before admission as midshipmen, candidates must deposit $1,500 in the Midshipmen's Store, to be used in partial payment for uniforms and supplies. In case of extreme hardship, this sum may be reduced or waived. If this is necessary, money allowances will be reduced until the individual's account reaches prescribed levels, and the midshipman's accrued savings will be less than the class average upon graduation. The initial issue and the additional uniforms, clothing, textbooks, personal computer, and expenses required during the first year are valued at over $4,000. The deposit made at the time of entry is supplemented by an entrance credit of $1,500, which is an interest-free loan made by the government to defray the cost of the uniforms and equipment required during the first year. Repayment of the indebtedness is accomplished by monthly deductions during the four years at the Academy.

Midshipmen have a choice of more than 70 extracurricular activities (ECAs) at the Academy. Almost all are organized and run by the midshipmen themselves. Extracurricular activities fall into several basic areas: athletic, professional, recreational, musical, religious, academic publications, and Brigade support. Midshipmen are encouraged to take part in as many activities as their interests and time allow. A sampling of ECAs includes: Judo, Bicycle Racing, Rugby, Scuba, Military Parachute Club, Drill Team, Sportsman Club, Drum and Bugle Corps, Choir, Bible Study, Forensics, Astronomy Club, Theatre, Civic activities, and Honor Societies. The ECAs are popular adjuncts to other programs offered in support of the Academy's mission.

Every midshipman is introduced to sailing at the Naval Academy. Beginning with Plebe Summer, midshipmen learn basic sailing in the Academy's knockabouts, sloops, and yawls. The program covers water safety, boat handling, sail techniques, crew work, and marlinspike seamanship. The Academy encourages the midshipmen to continue sailing activities throughout the four year program, and many opportunities are offered to participate competitively in sailing events or just to sail for recreation.

All service academies make very clear what is expected of future graduates. Their policies are very strict in the sense that violations are not tolerated. Any midshipman found guilty of Honor Concept violations or found to be dependent on drugs or trafficking in drugs, or of homosexual or immoral activity, is forced to leave the Academy.

You should be certain that you are sincerely interested in becoming an officer in the Navy before enrolling in the Academy. The reasons most often given by plebes who resign from the Academy are that they enrolled under parental pressure and that they failed to realize how demanding the program was. They were not able to adjust to the demands of the military and professional aspects of the training. They were not prepared for the regimen of a service academy. However, the overwhelming majority of graduates of the Naval Academy feel strongly that all the sacrifices and hard work were well worth it. They received an excellent education, have promising careers, and are able to move immediately into responsible positions. They are serving their country in worthwhile and important jobs.

The key is to be prepared before you go to Plebe Summer and know what will be expected of you. Talk to midshipmen attending the Academy, your high school counselors, and representatives of the Academy. A personal visit to the Academy can help in your decision. The Academy has several visitation programs, and someone from the Admissions Office is always available to talk to you. Parents are also welcome to visit the Academy. Obtain as much information as possible; write to:

Director, Candidate Guidance
United States Naval Academy
Annapolis, MD 21402-5018

You can also call tollfree: 1-800-638-9156.

Chapter V

The United States Air Force Academy

History

Air Service pilots who returned to the United States from World War I were convinced that only by having an autonomous position could the Air Service establish its own doctrines, develop and procure aircraft suitable to its missions, and command its own forces. The dream became a reality during World War II when the War Department created three coequal autonomous commands within the Army: Army Air Forces (AAF), Army Ground Forces, and Services of Supply (later redesignated Army Service Forces).

After World War II, leaders of the AAF were aware that its autonomous position would be terminated unless legislation were enacted to legalize its status. After many months of conferences and hearings, the National Security Act of 1947 became law. Title II established the United States Air Force and placed it on an equal status with the Army and the Navy. The oath of office for the first Secretary of the Air Force, W. Stuart Symington, was administered by Chief Justice Frederick M. Vinson on September 18, 1947. General Carl Spaatz became the first Chief of Staff, US Air Force, on September 26. And so began a new era of air power from a beginning only forty years earlier.

The establishment of a separate Air Academy to train officers for the specialized functions of the Air Force had been the goal of the military air leaders almost as long as the wish for a separate service, but it was not until January 1949 that the first step was taken. Secretary of Defense James V. Forrestal appointed a service academy board to examine the requirements for an air academy. The board chairman was Dr. Robert Stearns, President of the University of Colorado; General Dwight D. Eisenhower, then President of Columbia University, was vice chairman. After examining the facilities and capabilities of both the Military and Naval Academies, the board recommended that an Air Force Academy be established immediately. Following a long period of congressional committee hearings, the 83rd Congress passed a bill on March 29, 1954, signed by the President on April 1, establishing a separate Air Force Academy.

The Secretary of the Air Force, Harold Talbot, immediately appointed a committee to select a site for the Academy. The selection committee, composed of prominent civilians and senior military personnel, visited sites in twenty-two states and screened more than 400 proposed locations. Secretary Talbot eventually named the Colorado Springs site as the permanent location for the Air Force Academy for the following reasons: It was a quiet, isolated but accessible area that would be conducive to a good academic atmosphere; the area citizens had excellent relations with military personnel; the State of Colorado offered to donate one million dollars toward the purchase price of the land, and the area had good climatic conditions. Nearby Lowry Air Force Base in Denver was designated as the temporary site of the Academy. Construction was begun in 1955.

While construction was going on, the 4,000 applicants for the first class were screened to select the final group to be admitted. On July 11, 1955, these 306 were sworn in at Lowry Air Force Base in ceremonies befitting the dedication of the first new service academy in the United States in more than a century. Lieutenant General Hubert Harmon, who had been a member of the site selection committee, was recalled from retirement after thirty-eight years of active duty to become the first superintendent. The Academy curriculum was designed and perfected under his direction to meet the constantly changing developments of the Aerospace Age. The fundamentals and most recent findings of science were blended with the social sciences and the humanities to form a balanced program of education for future Air Force officers.

In the summer of 1958 Academy personnel moved from Lowry to the partially completed permanent home twelve miles north of Colorado Springs. When the Cadet Wing moved into the new dormitory in late August, the academic area was still under construction, although sufficiently completed to support the small Wing. Yet less than a year later, even before the first class graduated, the Academy was recognized as an accredited institution of higher learning by the North Central Association of Colleges and Secondary Schools. On June 3, 1959, the Academy commissioned its first officers.

Since the first class graduated, the Cadet Wing has grown to more than 4,400 cadets. Part of this expansion included the addition of women. President Gerald R. Ford signed the legislation authorizing the admission of women on October 7, 1975, and the first women graduated in 1980. The Air Force Academy today is in the forefront of scientific and aerospace institutions of learning. It is recognized as one of the leading technical institutions in the country.

Mission of the United States Air Force Academy

The mission of the Air Force Academy is to provide instruction and experience to all cadets so that they graduate with the knowledge and character essential to leadership and the motivation to become career officers in the United States Air Force.

As a minimum goal, the mission is accomplished by providing cadets with the following: the basic baccalaureate-level education in airmanship, related sciences, the humanities, and other broadening disciplines; a knowledge and appreciation of air power, its capabilities and limitations, and the role it plays in the defense of the nation; understanding of the Air Force missions of air warfare, air defense, and military space research; high ideals of individual integrity, patriotism, loyalty, and honor; and a sense of responsibility and dedication to selfless and honorable service. The program has three main aspects: military, academic, and athletic.

The military program is designed to develop in the cadet the moral character and qualities of leadership desired in an Air Force officer. It instills a deep-founded belief in national defense, pride in the Air Force, and inspiration to give one's best in a lifetime of service to one's country. It equips the cadet with the fundamental military knowledge and skills required of a junior officer and the professional military education for continued development leading to the highest command and staff positions. It motivates the cadet toward a career in the Aerospace Age and provides a foundation for future specialization in manned and unmanned aerospace systems.

The academic program provides general courses in the liberal arts and sciences that furnish a foundation for future development in any of the numerous career fields open to Air Force officers. It provides elective courses to meet the needs of the individual cadet in developing full academic potential. It motivates the cadet for continued educational development after graduation by self-instruction and attendance at graduate school. It also prepares the cadet to fulfill his or her intellectual duties as a citizen and a dedicated public servant in the Air Force.

The athletic program serves to develop desirable traits of character such as persistence, desire to win, and aggressiveness that are essential to leadership. It develops those qualities needed to perform physical tasks without undue strain. It develops a positive attitude toward physical fitness. It promotes the esprit de corps of the Cadet Wing through healthy competition and pride in an outstanding athletic program.

The Air Force is closely involved with the latest developments and advances in technology and space exploration. Consequently, the program is constantly modified to keep pace with the changing technological world, but the basic mission of preparing cadets to become successful professional Air Force officers remains constant.

Facilities

The Air Force Academy opened at its permanent location in 1958, the year heralded as the first of the Aerospace Age. The buildings and grounds are as new and modern as jets, missiles, and spacecraft. The Academy area contains approximately 18,000 acres, of which 10,000 acres are suitable for the construction of an airfield and additional buildings. The site is at an altitude that ranges from 6,200 feet to 7,400 feet. The spacious grounds backed by the Rampart Range of the Rocky Mountains on the west merge with the distant plains to the east, giving the Academy an atmosphere of expanding space horizons.

The main cadet area, parade ground, and athletic fields are in the northwest portion of the Academy site. The buildings where the cadets live and attend classes, named for famous Air Force leaders, are designed in contemporary architectural style featuring glass, aluminum, steel, and white marble, accented with bright mosaic tile. The Air Force Academy has the most impressive physical facilities of all the service academies.

Undoubtedly the most widely publicized structure of the Academy is the magnificant Cadet Chapel. Its striking design is in direct contrast with the horizontal lines of the other Academy buildings. Triangular aluminum sections rise 160 feet to form seventeen towering spires. The Cadet Chapel was envisioned as the crowning architectural feature of the Academy campus. Its design evolved out of a desire to achieve a structure symbolic of the religious aspects of a cadet's life and also to provide chapels respecting the individuality of religious faiths. It contains separate Protestant, Catholic, and Jewish chapels as well as an all-faith worship room.

The cadet dining hall, Mitchell Hall, is on the south side of the cadet area. It is an impressive building, one block square under a single roof, and large enough to seat the entire Cadet

Wing at one time. It is enclosed by glass walls on three sides, permitting a view of the beautiful Rocky Mountains. It was named for General William (Billy) Mitchell, pioneer of military aviation.

Cadet classes are conducted in the large academic building called Fairchild Hall. On the east side of the cadet area, it contains 45 science laboratories, 5 lecture halls, the Academy library, a dispensary, and faculty offices. The library occupies the largest portion of the north wing of Fairchild Hall, to the extent of 78,000 square feet of floor space. A soaring spiral staircase provides the main access to its three floors.

Social activities are held in Arnold Hall, named for General Henry H. (Hap) Arnold, the first Commanding General, Army Air Forces. The social center includes a ballroom, an auditorium, a bowling center, recreation rooms, lounges, snack bars, an information, ticket, and tour office, and a cadet graphics shop.

The Academy planetarium, in the west section of the cadet area, is used not only to teach celestial theory and star identification to the cadets but also to give lectures and demonstrations to the general public and members of other educational institutions. The modern multiple projector can accomplish realistic simulation of the movements of comets, meteor showers, the northern lights, constellations, sunrise and sunset, and planets in orbit. All stars visible to the naked eye can be duplicated and moved backward or forward to show their past or future positions.

The cadet gymnasium, directly north of Vandenberg Hall, has facilities for every kind of sports enthusiast. It contains two swimming pools (one Olympic-sized); many athletic courts and areas used for physical education classes and intramural and intercollegiate sports; rooms for fencing, wrestling, gymnastics, boxing, handball, squash, and judo; basketball courts, and a rifle and pistol range. Surrounding the gymnasium are football fields, soccer fields, baseball diamonds, volleyball courts, and tennis courts. The field house contains a multipurpose area utilized for indoor track meets, other sports practices and physical education classes, a 6,600-seat basketball court, and a 3,100-seat hockey arena.

The two cadet dormitories are Vandenberg Hall and Sijan Hall. Vandenberg Hall has 1,320 rooms, squadron areas, hobby shops, chaplains' offices, counseling offices, and a cadet store. Sijan Hall, an 830-room dormitory, was named for the late Captain Lance P. Sijan, Class of 1965, the Academy's first Medal of Honor winner. The cadet dental clinic and tailor shop are located in the dorm. Two cadets are assigned per room, and all personal necessities are available within the dormitory areas.

Harmon Hall, named after the first superintendent of the Academy, is the administrative building and houses the offices of the Superintendent and his staff. It is on the west side of the cadet area. Also on the grounds are the Academy Hospital,

A cadet squadron assembles for noon meal formation in front of the Honor Code wall at the Air Force Academy.

the Officers and Noncommissioned Officers Clubs, Visiting Officers' Quarters, two family housing areas with public schools, a community shopping center, the Academy Preparatory School, a supply and services area, a 3,500-foot airstrip, and a 2,100-foot glider strip. The Falcon Football Stadium and the Eisenhower Golf Course, one of the finest in the area, are also among the Academy's facilities. In the nearby mountains of the Rampart Range, Farish Memorial is a recreational area set aside for the cadets and Academy personnel. Jack's Valley, the primary cadet training area on the Academy grounds, provides field conditions for basic cadet training. In sum, the Air Force Academy has some of the finest facilities of any academic institution in the nation.

Entrance Criteria and Application Procedures

General eligibility requirements for entrance into the Air Force Academy are similar to those of other service academies. To qualify for appointment consideration, you must:

- be at least 17 years old;
- not yet have passed your 22nd birthday on July 1 of the year you enter the Academy;
- be a US citizen (foreign students authorized for admission are exempt from the citizenship requirement);
- be of high moral character;
- meet high leadership, academic, physical, and medical standards;
- be unmarried, with no dependents.

Air Force Academy preparation guidance stresses that the high admissions standards, the fierce competition for appointments, and the demands placed on the cadets at the Academy require thorough preparation. It takes a well-rounded program of leadership, academic, and athletic preparation to be one of the few who can meet the Academy's challenge. A recent *USA Today* study lists the Air Force Academy as the fourth most activities as possible as long as their academic program does not suffer. Cadets may follow interests first stimulated in the classroom such as language, computer, and geology clubs or such activities as chess, rugby, soccer, Cadet Glee Club, radio and TV, and cadet magazines. Many cadets step out of the West Point community to help in veterans' hospitals, juvenile correctional institutions, and young people's organizations. The Cadet Scoutmasters' Council works with local Boy Scout units and annually hosts a camporee. The Cadet Public Relations Council sponsors cadet appearances at junior and senior high schools, for military and civic organizations, and on radio and TV throughout the nation. All cadets compete in some form of athletics on the collegiate or intramural level.

Student government at West Point is the cadet chain of command under the auspices of the Corps of Cadets. Cadets manage the customs and tradition programs (with authority to discipline), the underclass military drill and training programs, and the Cadet Honor System. Cadets also manage most of the intramural athletic programs and other extracurricular functions. The student body is organized as a brigade and is divided into four regiments having 12 battalions and 36 companies; a cadet serves as Brigade Commander. The company is the hub of cadet life. Competition in athletics, parades, and military training is between companies. Long-lasting friendships, social interaction, and tutorial assistance are fostered within the company organization.

Academic, military, financial, and other types of personal counseling are available to cadets at all times. Apart from this professional counseling, cadets can always seek advice from their peers in the cadet chain of command. Legal assistance is available; advice on wills, contracts, taxes, and certain other private legal matters is provided at no cost. Cadets receive complete medical and dental care. Frequent examinations ensure continued excellent health. If hospitalization becomes necessary, cadets receive treatment in the well-equipped Keller Army Hospital at West Point.

The number of vacations (leaves) and the amount of free

time a cadet has depend on seniority as well as military performance. A senior receives about twice as many weekend leaves per semester as a junior. A plebe (freshman) leaves the Military Academy on only four weekends in addition to the Christmas holidays and authorized athletic, extracurricular activity, or cultural trips. Additional weekend passes may be awarded to cadets based upon individual or unit achievement. All cadets take Christmas and summer leaves. Upperclass cadets also take a spring leave. Progressively greater amounts of free time allow cadets to be part of the collegiate subculture at the Academy and to mix with students from nearby campuses.

In return for the cadet's five-year active duty obligation and three-year reserve obligation after graduation and commissioning, the government provides all tuition and room and board costs as well as any other educational fees. A cadet receives more than $6,500 per year. The cadet must pay for a personal computer, his uniforms, and textbooks from this amount. As part of the Regular Army, a cadet is entitled to salary and Army benefits.

If you are thinking about entering West Point, you must be prepared for total involvement in the program. You cannot get through the four years successfully without a lasting commitment to do your best. You must retain a positive attitude about what you are accomplishing and set your sights not only on getting through a day at a time but also on what you will achieve upon graduation. Obtain as much information about the Academy as you can before making your decision. Write for information to Director of Admissions, US Military Academy, West Point, NY 10996-1797. Attending West Point is a valuable experience that will serve you well the rest of your life.

Chapter IV

The United States Naval Academy

History

At the time of the founding of the United States Military Academy, Navy leaders were trying to interest Congress in the idea of establishing a similar school for the Navy. Such proposals were not favorably received at the time. Although a short-lived school was set up in 1803 at the Washington Navy Yard, it was not until 1838 that a one-year school, the Philadelphia Naval School, was established at the Naval Home, Philadelphia. This school was essentially for the administration of examinations for promotion, and attendance at its classes was entirely voluntary.

In 1845, George Bancroft, Secretary of the Navy under President Polk and a strong believer in the concept of training future naval officers ashore, established the Naval School at the old Army post, Fort Severn, at Annapolis, Maryland, where the Severn River empties into the Chesapeake Bay. The school was officially founded October 10, 1845, but it was not until the following year that appropriations were authorized for $28,000 for "repairs, improvements and instruction at Fort Severn, Annapolis."

The first superintendent of the Naval School, Commander Franklin Buchanan, instilled in the school the stability and high standards necessary to carry it through its crucial first five years. During those early years, many of the present organizations, customs, and traditions were founded. Each new candidate stood a trial period of one year ashore and six months at sea. If he displayed the required officer qualities during this period, he became a midshipman and served at sea for three more years. After another year ashore, he was commissioned lieutenant, US Navy. English, mathematics, geography, French, and Spanish were included in the curriculum. Lectures were given in chemistry, physics, and ordnance. The practice of dismissing and turning back those found deficient in academics was instituted in 1846, following the June final examinations.

On July 1, 1850, the Naval School became the United States Naval Academy, and the course of study was condensed to four academic years. Summer training cruises gave the midshipmen seagoing experience to augment classroom work. Thus, the forerunner of today's basic four-year curriculum and summer cruise program first appeared at the Naval Academy over 140 years ago.

The year 1861 was a sad one for the Academy when midshipmen of the North and South fell in at their last parade together. While loyalty to the Union was requested, it was not enforced. Schoolmates and classmates parted in deep sorrow, many to meet later at sea during the bitter struggle of the Civil War. Those who chose to remain loyal to the Union embarked in the USS *Constitution* on April 24, 1861, and sailed to New York and then to Newport, Rhode Island, where instruction was continued at Fort Adams. The Fourth Classmen (plebes) were quartered on the ship, while Third Classmen (youngsters) were berthed ashore. The two upper classes were commissioned at the outbreak of hostilities and assigned to the Union fleet. In the meantime, Fort Severn was again utilized as an Army post.

Under Vice Admiral David D. Porter, the Naval Academy was reopened on September 11, 1865. The USS *Constitution*, which returned the midshipmen to Annapolis, remained as the station ship until 1870. Admiral Porter supervised the repairs at the Academy necessitated by Civil War damage and also obtained funds for many new buildings. In 1883 the Naval Academy began commissioning Marine Corps officers.

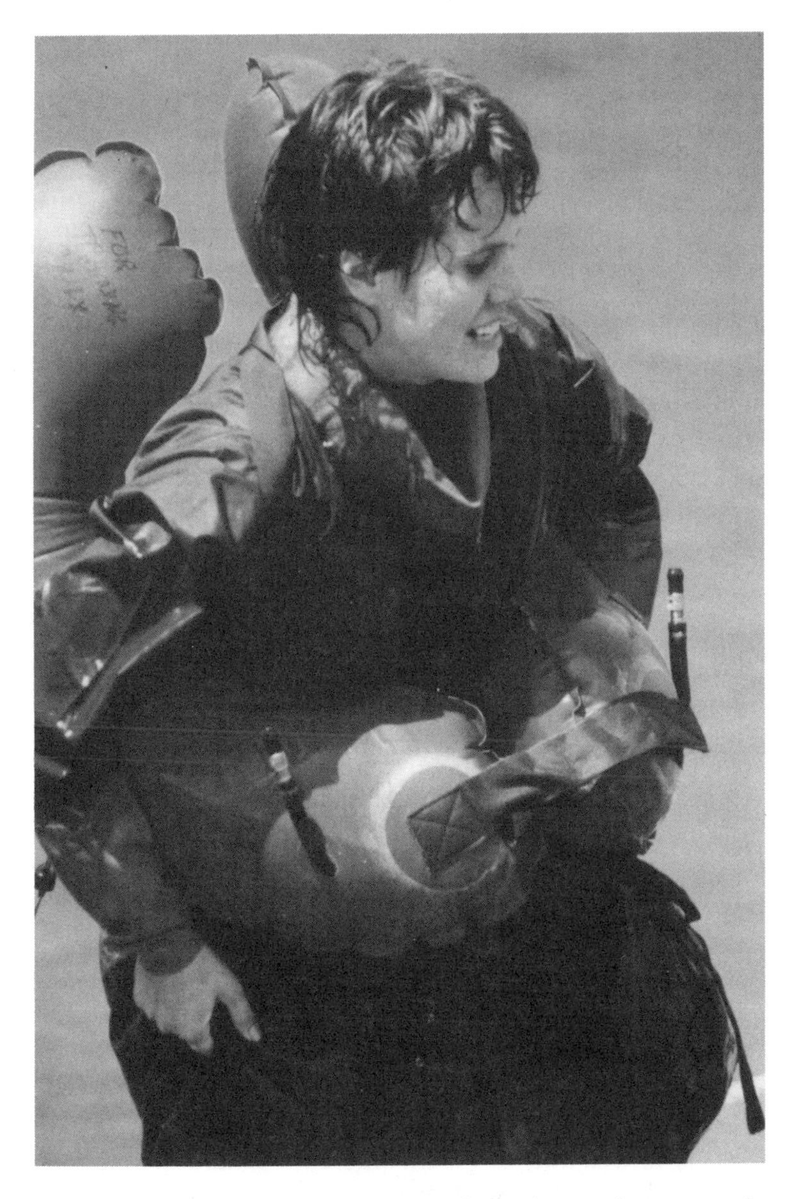

Water Survival is part of the SERE training during the second summer at the Academy.

selective college in the country. Consider carefully the characteristics of dedication, desire to serve others, ability to accept discipline, sense of duty and morality, and enjoyment of challenge in deciding whether you want to pursue an Air Force Academy education and a military career. The decision to accept this challenge must be *your own*. Do not let parents, friends, or others overly influence you. Outside influence, no matter how well intentioned, seldom provides motivation to meet the challenges you will encounter. You should start preparing early; junior high is not too soon. However, wait until after January 31 of your junior year to begin official application procedures.

A college preparatory high school education provides the best experience for the academic challenges of the Academy. Studies should include four years of English, including a college preparatory course in writing; four years of math, with a strong background in algebra, trigonometry, functional analysis, and analytic geometry; and four years of basic lab sciences. You should have two years of social sciences and two years of foreign language. A course in typing is helpful since cadets frequently prepare reports, themes, and research papers. A basic course in computers is helpful because you will be using your personal microcomputer during your four years at the Academy. Try to get the best grades possible. Plan to take the ACT or SAT tests more than once, since they provide a large portion of your academic composite. These tests have slightly different emphases, so it is advisable to take both tests to see which one measures your educational background most favorably. Also, work toward developing effective study habits and efficient time management, qualities that are essential in meeting the Academy's academic challenges.

Physical fitness is an integral part of Academy life. Part of the admissions process is the Candidate Fitness Test (CFT). The test consists of four events—pull-ups, push-ups, sit-ups, and 300-yard shuttle run—with different standards for men and women. Because the Academy experience makes intense physical demands, you should participate in individual and team sports throughout high school as well as an individual fitness program. Upper body strength, running speed, and endurance should be your objectives. You should practice swimming for endurance and learn survival techniques.

In the leadership portion of the whole-person composite, candidates are rated on a variety of athletic and nonathletic activities. Quality of involvement rather than quantity is the key. It is better to demonstrate leadership in a few selected activities than to be a "joiner" of many. The following list shows the percentages of high school extracurricular positions held by cadets in a typical Academy class:

Activity	*Percentage of Class*
Valedictorian	12
Class President	13
Boys/Girls State	25
Eagle Scouts	9
National Honor Society	73
Athletic Letter Award	81

The Academy evaluates candidates on the basis of numerous academic and nonacademic criteria that are combined to form a "whole-person" score. The academic component constitutes the major portion of the score. It includes SAT or ACT scores, with math counting more heavily than English, and the prior academic record including grades, rank in class, and any college courses taken. Honors and advanced placement courses receive extra credit. The SAT and ACT test scores, minimum and mid-50 percent range, of candidates appointed to the Academy in a recent class are shown in the following table:

	Minimum	*Mid-50 Percent Range*
SAT		
Verbal Aptitude	500	540–620
Math Aptitude	550	630–710
ACT		
English	21	23–27
Social Studies	19	24–29

Mathematics	24	27–32
Natural Sciences	24	28–32

Rank in Class

1st in Class	10%
Top Tenth	83%
Top Fifth	92%
Top Quarter	97%

It is wise to take one or both of these tests in your high school junior year or sooner. Then you may improve on previous scores by retaking the tests in your senior year. You should take one of them not later than November of your senior year to have your scores included in the final evaluation report that the Academy sends to members of Congress in January. When you register for the tests, request that your scores be sent to the Air Force Academy. The results of one of these tests must be in your records to be considered for an appointment. The SAT code number for the Academy is 4830, and the ACT code number is 0530. Consult your guidance counselor or college information office for registration procedures and test dates.

The extracurricular component makes up the remainder of the whole-person score. Basically, it measures athletic and nonathletic activities. Some of the athletic indicators are varsity letters earned, all-league or all-state recognition, or selection as a team captain. Nonathletic indicators include leadership in a club, participation in student government, involvement in Scouting or the Civil Air Patrol, or an after-school job. Performance on the Candidate Fitness Test is also included in the extracurricular component. By using these criteria, the Academy obtains candidates who can manage time and excel in the rigorous program. Competition for the Academy is keen; you should learn as much as possible about the selection procedures to give yourself the best chance for acceptance. Talk to someone who has gone through the process if possible, as well as counselors, Academy officials, and regional Air Force liaison officers.

The first step toward becoming a cadet is to request a Precandidate Questionnaire (PCQ) from the Admissions Office. The Academy uses the PCQ to accomplish three purposes:

- evaluate the qualifications of prospective candidates;
- provide information on potential nominees to members of Congress;
- schedule those who qualify for medical examinations.

To obtain a PCQ, write to the Admissions Office, HQ USAFA/RRS, USAF Academy, Colorado Springs CO 80840-5651. To be sure your name is included in reports to members of Congress, return the questionnaire as soon as possible. Be sure to include your social security number. Your completed application must be postmarked by January 31 of the year you intend to enter the Academy. However, most applicants return the questionnaires by the end of the previous October. If you need assistance, talk to your counselor or your nearest liaison officer.

If your PCQ information indicates that you do not meet the minimum qualifying levels, you will be notified. You can try to qualify later by submitting higher scores or more current information. If you meet the qualifying levels, you can focus your efforts on obtaining a nomination. You may apply to the Academy each year until you are accepted or until you exceed the age limit.

Nomination Categories

Nomination procedures for the Air Force Academy are similar to those for the Military and Naval Academies. No political affiliation is necessary to apply for a nomination. Members of Congress want to nominate outstanding persons who are competitive for the service academies. Since the process can be lengthy, you should start seeking your nomination at the same time you return your PCQ. To increase your chances of being selected, you should apply for a nomination in each category in which you are eligible. You should write to your US Representative, your two US Senators, and the Vice President. You may also be eligible in various military-affiliated categories as listed below. Sample formats for these letters are given in the Appendix. The deadline for requesting a nomination in the Vice Presidential category is October 31. All other categories have an Academy deadline of January 31. However, many members of Congress do not accept requests for nomination after October, so you should contact them in the spring of your junior year.

The Congress is the primary source of nominations. Senators and Representatives have considerable latitude in awarding nominations. Most awards are based on some combination of academic achievement, leadership performance, and athletic participation. Each member of Congress may have a maximum of five cadets attending the Academy at any one time. When one of those cadets graduates or is disenrolled before graduation, a vacancy exists. For each vacancy, the member of Congress may nominate a maximum of ten candidates. These ten candidates are ranked according to qualification by the member of Congress, or they are ranked by the Academy based on best-qualified criteria. The same methods of nominating candidates are also available to the Vice President (who nominates from the US at large), the delegates to the House of Representatives from the District of Columbia, Guam, the US Virgin Islands, and American Samoa; the Panama Canal Commission Administrator; and the Resident Commissioner and Governor of Puerto Rico.

If you are eligible to apply in a military-affiliated category, you can use the sample letter given in the Appendix. Mail the letter to Director of Admissions, HQ USAFA/RRS, USAFA, Colorado Springs, CO 80840-5651. The categories are as follows:

Presidential. This category is reserved for children of career military personnel. To qualify, the parent must meet one of the following criteria:

- Be on active duty and have served continuously on active duty for at least eight years;
- Have retired with pay or have been granted retired or retainer pay. Children of reservists who retired while not on active duty are not eligible;
- Have died after being retired with pay or being granted retired or retainer pay.

Children of deceased or disabled veterans of military or civilian personnel in a missing status. The child of a deceased or disabled member of the Armed Forces is eligible if the parent was killed or 100 percent disabled by wounds or injuries received or diseases contracted in active service or from a preexisting injury or disease aggravated by active service. The child of a parent who is in "missing status" is eligible if the parent is a member of the armed services or a civilian employee in active government service who is officially carried or determined to be absent in a status of missing; missing in action; interred in a foreign country; captured, beleaguered, or besieged by a hostile force; or detained in a foreign country against his or her will.

Children of Medal of Honor recipients. The children of Medal of Honor recipients from any branch of the armed services may apply under this category.

Air Force Regular and Reserve components. Nominations are available under this category for members of the Regular Air Force, Air Force Reserve, and Air National Guard in accordance with Air Force Regulation 53-10. The application form (AF Form 1786) can be obtained through Air Force publication supply channels.

Air Force Reserve Officers Training Corps. Five students may be nominated to the Academy every year from each college or university AFROTC detachment as well as from each high school with Air Force Junior ROTC. College or univeristy students must submit their applications to the Professor of Aerospace Studies. High school students must submit their applications to the Aerospace Science Instructor.

Honor military and naval schools. Any school designated by the Departments of the Army and Navy as an honor school may nominate five candidates from among its honor graduates. Applications are provided by the Academy to eligible schools.

The best advice on admissions, as stated previously, is to get an early start on applying. Seek out all the information you can obtain about the Academy from the Admissions Office. Talk to former and current cadets if possible and to Admission Liaison Officers in your region. The Academy Admissions Office can supply the name of the Liaison Officer. Most of all, make sure it is *your* decision to apply to the Academy.

Academic and Military Programs

The academic program, under the direction of the Dean of the Faculty, allows cadets to acquire a broad education in the basic sciences, engineering, humanities, and the social sciences. Cadets must complete a balanced sequence of prescribed courses in all areas. Beyond the prescribed courses, they may choose a major in a specific area, or they may elect the basic academic program of study. All choices lead to a bachelor of science degree if successfully completed. Each cadet must earn at least 176 semester hours of credit. This requirement is greater than those at civilian universities.

The Academy education program continues year round and is divided into three sessions: a summer term, a fall semester, and a spring semester. The summer term lasts approximately ten weeks. The new cadet class usually enters the Academy during the first week in July, the midpoint of this term, and begins Basic Cadet Training (BCT). During the summer term, the three upper classes participate in leadership and military training programs at the Academy and at military bases around the United States and abroad. A limited number of academic courses is also offered. All cadets except the new class receive about three weeks of summer leave. The fall and spring semesters contain approximately 17 weeks of instruction, or 42 lessons per semester. The fall semester begins in mid-August and ends the week before Christmas. The spring semester begins the first week in January and ends in mid-May. Each semester includes a final examination period of six days.

All cadets must meet the following graduation requirements:

- Demonstrate an aptitude for commissioned service and leadership by having a minimum cumulative military performance average (MPA) of 2.0.
- Be satisfactory in conduct.
- Be proficient in physical education and military training.
- Complete the requirements for the core curriculum and for an academic major or the basic academic program.
- Have a minimum cumulative grade point average (GPA) of 2.0.

The course of instruction is built around prescribed core courses in the first and second years and mostly elective courses with some required courses in the third and fourth years. The summary of the curriculum on page 100 shows the standard sequence required of most cadets. In a two-course sequence, the first course is offered in the fall. Single-course offerings are split between fall and spring to help balance departmental workloads. Cadets with advanced standing take some courses ahead of schedule.

Cadets do have a choice in the shape of their curriculum. They can choose the basic academic program or choose to major in one of 25 specific subject areas. The basic academic program provides a broad educational foundation with maximum flexibility in selecting advanced courses. In addition to the core curriculum, 11 academic divisional optional courses and two open option courses are taken. The basic academic program is not a major, but graduates do receive the bachelor of science degree. For a major in a specific subject area, the

Summary of the Curriculum

Class of 1992

Fourthclass (Freshman)

Summer

Course Title	Academic Hours
Basic Cadet Training	
Basic Physical Training	

Fall and Spring

Course Title	Academic Hours
General Psychology	3
Principles of Chemistry	5
Introduction of Computer Science	3
Fundamentals of Engineering Mechanics	3
Language and Expression I	3
Basic Foreign Language	4.5
Modern World History	3
Calculus I and II	6
General Physics	3
Professional Foundation	1
US Defense Establishment	1
Introduction to Aviation Fundamentals and Space Science	.5
Flight Fundamentals	.5
Physical Education	3

Thirdclass (Sophomore)

Summer

Course Title	Academic Hours
Survival, Evasion, Resistance and Escape	
CONUS Field Trip	
Soar for All or Basic Parachuting*	

Fall and Spring

Course Title	Academic Hours
Introductory Biology	3-4
Principles of Microeconomics	3
Language and Expression II	3
Introduction to Military History	3
Introduction to Management	3
Probability and Statistics	3
General Physics II	3
Politics and American Government	3
International Politics/National Security	3
Airpower Theory and Doctrine	1
Soar for All	2
Glider or Parachuting Instruction	2
Physical Education	3

Secondclass (Junior)

Summer

Course Title	Academic Hours
Operation Air Force	
Summer Military Options	

Fall and Spring

Course Title	Academic Hours
Fundamentals of Aeronautics	3
Introduction to Astronautics	3
Leadership Concepts and Applications	3
Base Comprehensive Planning and Design	3
Principles of Macroeconomics	3
Electrical Signals and Systems	3
Energy Systems	3
Major Works of Western Literature	3
The American Legal System	3
Ethics	3
Academic Elective Courses	12-24
United States Force Employment Concepts	2
Physical Education	3

Firstclass (Senior)

Summer

Course Title	Academic Hours
Summer Military Options and Flight Core	

Fall and Spring

Course Title	Academic Hours
Engineering Systems Design	3
Law for Commanders	3
Academic Elective Courses	27-30
Military Theory and Force Analysis	3
Physical Education	3
Flight Core	3

* May be taken during the thirdclass summer or thirdclass academic year.

Flight core may be taken summer, fall or spring. Total hours depend on major elected (if any) and flight core scheduling.

requirements for graduation are 30 core courses plus 15 major courses for a divisional major, or 16–18 major courses for a disciplinary or interdisciplinary major. Cadets can choose an academic plan after successful completion of two or three semesters taking the prescribed courses. A faculty adviser explains the requirements of all majors and assists in planning a suitable course program based on a major or nonmajor track. Faculty departments and divisions administer the following 25 academic majors:

DISCIPLINARY MAJORS

Science and Engineering

Aeronautical Engineering
Astronautical Engineering
Biology
Chemistry
Civil Engineering
Computer Science
Electrical Engineering
Engineering Mechanics
Engineering Sciences
Mathematical Sciences
Physics

Social Sciences and Humanities

Behavioral Sciences
Economics
English
Geography
History
International Affairs
Legal Studies
Management

INTERDISCIPLINARY MAJORS

Operations Research
Space Operations

Falconry is one of the many extracurricular activities at the United States Air Force Academy.

DIVISIONAL MAJORS

Basic Sciences
General Engineering
Humanities
Social Sciences

A cadet who majors in a subject area must complete all the requirements. A major may be changed if the new requirements can be met without excess overload courses. It is possible to earn more than one major if all requirements are fulfilled. You must maintain acceptable standards in your academic, military, and physical education performances.

Military Training

The cadet's first exposure to military life occurs in Basic Cadet Training (BCT), a rigorous orientation program during the first summer at the Academy. When you arrive, you realize immediately that the Air Force Academy is not a typical college or university—it is a military institution. Your civilian clothes are turned in for military uniforms, and your hair is cut to military standards. Smoking is not allowed during BCT. If you smoke, you should try to quit before you arrive so you can adapt more easily. Waiting to give up smoking until you arrive at BCT is a very bad course of action. The five-week BCT consists of two phases: one in the Cadet area, and the other in Jack's Valley.

The BCT in the Cadet area focuses on the transition from civilian to military life. Upperclass cadets instruct the new students in military topics ranging from customs, courtesies, and the Air Force heritage to rifle manual of arms and marching. The new students are given plenty of opportunity to demonstrate proficiency through knowledge tests, drill and rifle manual competitions, parades, and inspections. They also fly in several different aircraft and complete the obstacle course. Physical conditioning is an extremely important part of BCT. The daily training includes exercises, running, and competitive sports. All activities are designed to condition the cadet for the physical demands that come during the school year.

Following the military and physical preparation of BCT in the Cadet area, training continues in the field setting of Jack's Valley. While in the field environment, cadets wear the fatigue utility uniform with combat boots and live in tents. They take part in many activities designed to test their physical limits and build confidence in their physical abilities. They also receive instruction in small-unit tactics, firearms, and air base defense. The military training begun in BCT continues throughout the four years at the Academy. It is an integral part of the summer and academic year programs.

Fourthclass Year (Freshman). The first military training experiences in BCT prepare new students to become part of the Cadet Wing. The Professional Military Training (PMT) program provides the initial foundation of Air Force knowledge and skills necessary to make the transition into the Academy environment and the leadership skills needed to achieve success in the military profession. Fourthclass PMT concentrates on developing an appreciation for the qualities of a professional officer, particularly the need for building individual professional values. Cadets spend considerable time studying regulations, the Air Force Cadet Handbook, and Academy heritage.

Thirdclass Summer. All cadets take SERE (Survival, Evasion, Resistance, and Escape) training during the thirdclass summer. The training is conducted at the Academy and in the nearby Rocky Mountains. During the one-week Continental United States (CONUS) field trip, cadets make short visits to two Air Force bases to become acquainted with the operational missions of the major air commands. They also participate in either the parachuting or soaring programs. In parachuting cadets train in free-fall parachuting and become familiar with emergency parachuting as it pertains to Air Force careers. After successful completion of the program, they receive the parachutist badge. The soaring program provides firsthand experience in flying through instruction in ground school and flights in sailplanes.

Thirdclass Year. The thirdclass PMT program builds upon the foundation already established. Thirdclass PMT concentrates on the transition from follower to leader. Cadets receive instruction in effective leadership techniques to enable them to lead and train fourthclass cadets during the next two years.

Secondclass Summer. The third summer at the Academy emphasizes leadership application and increased knowledge of the Air Force. Cadets may serve as instructors in BCT, SERE, parachuting, soaring, navigation, and other programs. They participate in Operation Air Force, a three-week worldwide program at an operational Air Force base, to observe and gain better insight into the duties of officers and enlisted personnel.

Secondclass Year. The secondclass PMT program stresses leadership and supervisory skills for application in increasingly responsible upperclass leadership roles. It provides opportunities for cadets to assess and improve leadership abilities through small-unit leadership situations. Secondclass cadets also receive additional instruction in honor and human relations issues. These discussions focus on developing professional officers with strong character and a mature sense of responsibility.

Firstclass Summer. During the fourth summer, cadets refine their leadership skills and prepare for commissioning as second lieutenants. They assume senior leadership positions in such programs as BCT and SERE training. Several optional programs are available to round out summer training requirements. Aviation is required of all cadets not scheduled to enter undergraduate pilot training. Academics, trainers, simulators, and flight experience in T-43 jet aircraft provide knowledge of aviation fundamentals and a familiarity with the aircrew environment. The aviation instructor program offers selected cadets technical and professional training to serve as aviation instructors during the academic year. Military training options provide opportunities for cadets to select programs that will help round out their professional development.

Firstclass Year. The firstclass PMT program concludes cadet military training at the Air Force Academy. It focuses on the ability to apply previously learned leadership and management principles, while providing the practical knowledge needed for entry into the Air Force.

Professional Military Studies

The Commandant's program in Professional Military Studies (PMS) focuses squarely on the profession of arms. The PMS curriculum stresses military professionalism, officership, the art of war, air-power heritage, military theory, doctrine, and force employment. PMS provides solid experience in the traditions, tools, operating practices, theoretical principles, and challenges of the military profession. Core courses in PMS are required for each of the four years at the Academy. The first course, Professional Foundations, expands the honor training received in BCT. It examines the professional roots of the military and the tradition of the Air Force and shows how integrity, honesty, and moral courage are part of officership.

The next PMS course, US Defense Establishment, studies the basic mission, organization, and operation of the US Armed Forces, particularly the concept of air power and the Air Force role in supporting national objectives. It provides a general understanding of the current defense posture.

The thirdclass PMS course focuses on issues vital to the war-fighting role of the Air Force. Building on the knowledge acquired in the first PMS courses, the cadet studies basic airpower doctrine and employment concepts. Case studies are used to examine how air commanders have applied the principles of war in specific battles and campaigns. A war game allows cadets to apply the principles of war and practice military leadership in a simulated combat environment.

The secondclass PMS course provides an understanding of basic service doctrine and its relationship to current force structure. It also examines how the four branches of the armed services apply current employment concepts on the modern battlefield. A war-game exercise helps to clarify the purposes and complexities of joint staff planning.

Firstclass cadets study the writings of significant military theorists and examine the military force structures of selected foreign nations. Although the firstclass PMS course places special emphasis on the military forces of the Soviet Union and the Warsaw Pact nations, it employs analytical techniques that can be used as a basis for studying the military capabilities of other nations as well. This foundation in military theory and force analysis serves the officer well throughout an Air Force career.

Opportunities are numerous to receive aviation instruction and participate in flying activities. Cadets may earn Federal Aviation Administration (FAA) ratings in various aircraft and soaring training. Extracurricular flying as an activity is available.

Cadet Life

Life at the service academies is not like life at civilian institutions. It is a military environment, and the military permeates all aspects of life at the Academy. The structure, rules, regulations, and schedules help cadets organize their time and establish the self-discipline necessary to meet all the challenges offered. The academic schedule is similar to that of a civilian university, with four 50-minute periods each morning and three each afternoon. Cadets form up for reveille, march to lunch, and after classes they join in athletic activities. Those who do not participate in intercollegiate athletics play on a squadron intramural team two afternoons a week. The other three afternoons are unscheduled and may be used for study or for personal business. Intercollegiate athletes usually practice or compete every afternoon and frequently on weekends. Many cadets take additional academic instruction during the hour after classes or during other unscheduled times. Cadets spend many evenings studying in their room or in the library. They must be in bed at taps unless they have permission to study late. The following schedule depicts typical weekday activities during the fall and spring semesters. Saturday mornings are often devoted to parades, inspections, and study. Cadets are usually free from duty on Saturday afternoons and Sundays.

Typical Daily Schedule

6:00	Release from night call to quarters
6:30	Reveille
6:30–6:55	Early breakfast for cadets with graded review period
6:55	Day call to quarters
6:55–7:50	Common graded review
7:10	Squadron assembly and parade
7:30	Breakfast
8:00–11:50	Academic classes or individual study time
12:05	Lunch assembly and parade
12:25–12:50	Lunch
1:00–3:50	Academic classes, military training or individual study time
3:00–6:30	Intercollegiate practice
4:20–6:20	Intramurals or unscheduled time
4:30	Retreat (Fridays, or as scheduled)
5:00–7:45	Dinner
7:00–7:50	Optional evening lecture
8:00–11:00	Study
11:00	Taps

Time away from the Academy during the first year is very limited. During the first five weeks, in BCT, cadets may not have visitors, leave the Academy, or receive phone calls. After BCT the restrictions relax somewhat; cadets may receive phone calls and have visitors on Saturday afternoons and evenings and on Sunday mornings and afternoons. They may leave the Academy and go into the local area several times a month. Occasionally, they may be invited to dinner in the homes of Academy personnel. Throughout the four years, cadets are required to attend home football games and other selected Cadet Wing events.

Freedom increases gradually in the upperclasses. Thirdclass (sophomore) cadets still have limited free time. The second-class (junior) year offers more free time. Firstclass (senior) cadets have every weekend free except when military duties such as parades, training, inspections, or football games are scheduled. Most cadets go to Colorado Springs during off-duty time. You may not own or maintain an automobile as a fourth- or third-class cadet but may rent one while on an authorized pass or privilege. Second- and firstclass cadets may own cars and keep them at the Academy. On a Friday or Saturday pass, cadets must return to the dormitory by 1:30 a.m. On a weekend pass they may remain away from the Academy following the last military duty Friday or Saturday until Sunday evening study time.

Cadets receive more or fewer passes depending on individual achievement or deficiency. If performance is not satisfactory in military training or academic studies, free time may be restricted. Cadets receive about three weeks of summer leave except for the summer they enter the Academy. Christmas leave is about two weeks, and spring leave, one week. Emergency leave may be granted if an emergency involves a member of the cadet's immediate family. Other requests for special leave are considered on an individual basis.

All aspects of cadet life center around the Cadet Wing, which is organized on a military chain of command. Each cadet is assigned to one of 40 squadrons of 110 cadets. The squadrons are organized into four groups. Upperclass cadets learn leadership skills by filling command and staff positions within this structure. They lead the Cadet Wing not only during the school year, but during the summer as well. Such summer programs as BCT, SERE, and the various aviation courses are run almost entirely by cadets, who serve as commanders, instructors, and staff. At the top of the military hierarchy of the Cadet Wing is an officer chain of command. The Commandant of Cadets, a brigadier general, is the commander of the Cadet Wing. His staff of officers control the Cadet Wing, military instruction, and cadet aviation programs and advise the cadets on effective leadership of the Wing. The cadet and officer chains of command interact to provide cadets the opportunity for input to the policies and regulations that direct their lives. Both focus on maintaining the military atmosphere and discipline necessary for the proper development of future officers.

One of the most important features of cadet life at the Academy is the Honor Code. All cadets must abide by the words "We will not lie, steal, or cheat nor tolerate among us anyone who does." The Honor Code is specific and clear in its demands. Cadets are expected to have complete integrity in both word and deed; they will not lie or quibble, and they are expected to report themselves for any Honor Code violation. They are also expected to confront any other cadet who they believe has violated the Code and to report the incident if their suspicions remain. The Honor Code is not a difficult standard to live by. Initially it requires some self-control and conscious effort, but it quickly becomes an ingrained habit and part of one's total behavior. The Honor Code greatly enhances the quality of life at the Academy. It is a comforting feeling to be able to rely completely on the word and integrity of all one's fellow students. Cadets and graduates alike cherish the experience. The Honor code provides the basis of the professional ethics that officers must possess for military service. Breaches of the Honor Code are considered serious matters and can result in disenrollment.

A multitude of extracurricular activities are available during the four years at the Academy. Over the years, cadets have originated and continued more than 75 professional, recreational, and competitive activities. Some activities provide opportunities for weekend trips and competition with regional or national teams and the other service academies. They range from Choir, Scuba Club, Chess Club, Karate Club, Drum and Bugle Corps, Magic Club, and Yearbook to hunting and fishing clubs; there is something for everyone. Religious activities, counseling and advising, medical and dental care, and legal services are available to all cadets.

All cadets receive education, room and board, and monthly pay while at the Academy. All cadets are paid a gross salary of over $500 per month. During your fourthclass year, when you make an initial deposit of $1,000, you will receive a $60 cash allowance per month, after all expenses have been deducted. In your third-, second-, and firstclass years, your cash allowance will increase because of lowered required deductions. Required deductions include federal income tax and social security tax. Personal service deductions include barber, tailoring, linen, laundry and dry cleaning charges. Other deductions are for uniforms, textbooks, personal computer, publications, and membership fees. You may also elect to have life insurance and charity contributions deducted from your pay. If the cadet is economical, pay and allowances are adequate for self-support and include a provision for savings. Cadets receive these savings upon graduation to help purchase uniforms and meet other initial expenses as an officer.

Many exciting and challenging opportunities are available to the graduating cadet. You accept an eight-year obligation upon graduation. After five years of active duty you become eligible to request a separation from the Regular Air Force and serve the remaining three years in the inactive Reserve. About 65 percent of the graduates enter pilot or navigator training, which incurs an additional service commitment.

Find out as much as you can about the Academy before you make your decision. If possible, visit the campus for a tour and talk with officers in the Admissions Office. The Academy has Admissions Liaison Officers throughout the country who can assist you and provide information about the Academy. You can obtain the name of your closest Liaison Officer from the Admissions Office, USAFA/RRS, USAF Academy, Colorado Springs, CO 80840-5651. Telephone numbers for the Academy are in the Appendix.

Chapter VI

The United States Coast Guard Academy

History

Small in number of students, but large in scope of activities, the United States Coast Guard Academy was established in 1876 to provide career officers for the nation's oldest armed force afloat. The United States Coast Guard is unique among the military services. In peacetime, functioning under the Department of Transportation, it is our foremost agency for the advancement of maritime safety. Its mission is essentially humanitarian. In time of war, however, or by the direction of the President, it becomes part of the United States Navy.

Its history began in 1790 when Alexander Hamilton founded the Revenue Marine for the suppression of smuggling and the collection of tariffs. The first of the ten boats authorized by Congress was the *Massachusetts*, launched in 1791. It was a fifty-footer, manned by a crew of eight and armed with six small guns. By 1863 the modest little fleet of 1790 had come a long way. Treasury Secretary Salmon P. Chase had twenty-eight cutters in the Revenue Marine upon which he could call. The Revenue Marine of 1863 was, however, a loosely organized agency, with its vessels under the jurisdiction of various collectors of customs. During the Civil War the cutters carried out their traditional tasks of assisting vessels in distress and protecting private shipping.

In 1876 the Congress provided for the appointment of cadets to fill the lower commissioned ranks of the Revenue Cutter Service, as it was then known. In that year the first Coast

Guard Academy class, numbering nine cadets, boarded the old revenue cutter *Dobbin* at Baltimore and the Academy was officially established. The *Dobbin* was a two-masted topsail schooner and the first of a line of floating schools. Two years later, the slightly enlarged Corps was using the modern 250-ton bark *Chase* as its school. Named for Lincoln's Secretary of the Treasury, the vessel was bark-rigged, 106½ feet long, with a 25½ foot beam, and armed with four-inch guns. The *Chase* served as the cadet training ship until 1907.

For economic reasons the Academy was inactivated for three years beginning in 1890. Officers for the Revenue Service were drawn from the graduating classes of the US Naval Academy at Annapolis during that period. In 1900 the Coast Guard Academy was moved to a land base in Curtis Bay, Maryland. In 1907 the *Chase* was replaced by the *Itasca*, a barkentine-rigged steamer built by the Navy as the new cadet training ship. This vessel was not considered entirely successful, principally because she was scaled down in all dimensions including hatch sizes, bunk lengths, and overhead clearances. The Academy was relocated again in 1910. Sites considered were Portsmouth, New Hampshire; Port Royal, South Carolina; and New London, Connecticut. The Connecticut site was chosen, and the Academy was established at the Revolutionary War fort and old Army post, Fort Trumbull.

In 1914 the *Itasca* was replaced as the training ship by the *Alexander Hamilton*. The outbreak of World War I caused drastic changes in the Academy's curriculum, and cadets were required to train naval recruits in addition to working for their own commissions. The demand for officers grew steadily as the war continued, and in 1918 two classes entered the Academy simultaneously.

In the years before 1926 the Academy offered few if any engineering courses, but in that year the distinction between an engineering and a line officer was dropped, and the present-day multifaceted course of instruction was adopted. By 1932 the Academy had expanded to such an extent that larger accommodations were required. Land was donated to the Academy, and new buildings were constructed. These buildings and their subsequent additions comprise the present facilities.

At the start of World War II classes at the Academy were accelerated, and the class of 1942 was graduated six months ahead of schedule. In 1948 the four-year course was reinstated and a postwar curriculum established. The present Academy curriculum, a far cry from the training given aboard the *Dobbin* in 1876, is a well-rounded and carefully integrated balance of cultural, scientific, engineering, and professional subjects. From the original class of nine cadets appointed by Secretary of the Treasury John Sherman, the Coast Guard Academy has grown in numbers and facilities until today it counts an average enrollment of 300 new cadets every year and is located on a 120-acre reservation.

While the Academy has changed its location a number of times in its history, two things have not changed. First, from the beginning cadets were selected solely on the basis of merit; and second, honor, courage, devotion to duty, and high standards of professionalism have always been emphasized.

Mission

The United States Coast Guard Academy at New London operates under the general direction and supervision of the Coast Guard. Its purpose is to provide commissioned officers for the United States Coast Guard, a military service under the Department of Transportation in time of peace and under the United States Navy in time of war.

Through a four-year program of training and education on the college level, the Coast Guard Academy fulfills its stated mission: "To graduate young men and women with sound bodies, stout hearts, and alert minds, with a liking for the sea and its lore, and with that high sense of honor, loyalty and obedience which goes with trained initiative and leadership; well grounded in seamanship, the sciences, and the amenities, and strong in the resolve to be worthy of the traditions of the commissioned Officers of the United States Coast Guard in the service of their country and humanity."

The accomplishment of this mission is achieved through a curriculum composed of an academic program, a summer training program, and a physical education and athletic activities program. Successful completion of the curriculum entitles the cadet to graduate with a bachelor of science degree and a commission as an ensign in the United States Coast Guard.

The academic program is designed to provide the cadet with a fundamental education that will qualify him or her as an officer, to take an assignment in the Coast Guard and quickly become a productive member of the service. The subjects provide a thorough grounding in the physical and engineering sciences, the humanities, social studies, and certain essential seafaring and professional sciences. The course also provides a foundation for advanced study in graduate work if the officer has the aptitude and desire.

The summer training cruises are in effect a summer academic term. Cadets are instructed and graded in seamanship, gunnery, watch standing, and engineering. Most important is the determination of the cadet's adaptability to the service. The physical education and athletic activities program is designed to provide maximum development of strength, endurance, agility, and the basic physical skills. Throughout the program, every attempt is made to give the cadet opportunities to develop qualities of moral and physical courage, resourcefulness, group loyalty, fair play, leadership, and quick thinking while participating in highly competitive situations.

The mission of the Coast Guard has expanded significantly since the early beginnings. Today the Coast Guard has a variety of peacetime missions as well as maintaining defense readiness in case of war. Nearly 38,000 active duty men and women are stationed worldwide supporting the wide variety of Coast Guard missions. There are three primary areas: maritime law enforcement, safety of life at sea, and defense readiness. These areas are supported by twelve operating programs: aids to navigation, boating safety, defense operations, environmental response, ice operations, law enforcement, marine inspection, marine licensing, marine science, port safety and security, and search and rescue.

There is no question that the service is demanding; but, to the well-motivated, well-prepared officer who thrives on the age-old challenge of men and ships against the sea, it is an exciting and meaningful career.

Facilities

Housed in modern red brick buildings of New England Colonial design on the western bank of the Thames River, on a 120-acre site, is the United States Coast Guard Academy. Entering the main gate of the Academy, one is immediately confronted with a beautiful half-moon-shaped parade ground, Washington Parade. Fronting on the parade ground are three of the Academy's most important buildings: Hamilton Hall, Chase Hall, and Satterlee Hall. Hamilton Hall, the center of the three, is named for the first Secretary of the Treasury, Alexander Hamilton, and is the main administration building. Only within the past twenty-five years has the Coast Guard Academy come under the authority of the Department of Transportation; before that it was under the Department of the Treasury. Chase Hall, named for Salmon P. Chase, is to the left of Hamilton Hall and contains the cadet quarters. Satterlee Hall, to the right of Hamilton Hall, was named in honor of Captain Charles Satterlee, who lost his life aboard the cutter *Tampa* when she was torpedoed during World War I. It contains classrooms and a modern computer center.

The Academy operates under an ongoing modernization and improvement program. Behind Hamilton Hall is McAllister Hall, the engineering building, which contains the electronics, electrical engineering, and materials laboratories. Nearby is Michel Hall, a complete up-to-date hospital. Yeaton Hall, behind Satterlee Hall, is a professional studies building that contains a $5.1 million dollar ship's bridge simulator as well as classrooms and laboratories for ordnance and gunnery instruction. Down the hill from Yeaton Hall is Roland Hall, a field house and gymnasium that is one of the finest athletic facilities in New England. Billard Hall, the gymnasium, is at the south end of Cadet Memorial Field. It is named for Rear Admiral F.C. Billard, Commandant of the Coast Guard Academy from 1924 to 1932, who was an outstanding proponent of competitive sports as applied to the service. On the Thames River are the waterfront facilities including Pine Hall, the Seamanship Sailing Center, and the berth of one of the training ships, the *Eagle*. Adjacent to the waterfront is Nelson W. Nitchman Field, with baseball, softball, soccer, and track areas and a crew rowing center.

The beautiful Memorial Chapel was constructed with funds donated to the Coast Guard. Situated on the peak of the hill, the highest point of the Academy grounds, the white marble spire stands out for miles around as a monument in memory of the brave men and women of the Coast Guard who have died "in the service of their country and humanity." The chapel is nonsectarian. Overlooking the crew rowing center is Munro Hall, the enlisted men's barracks. It was named for Signalman First Class Douglas Munro, who died at Guadacanal in World War II while engaged in the evacuation of Marines from the beaches. Reportedly, his last words were, "Did they get off?" He was awarded the Congressional Medal of Honor posthumously.

Other buildings include Dimick Hall Auditorium, Smith Hall, the science building, and Leamy Hall, the student union and auditorium. The library, together with the US Coast Guard Museum, is housed in Waesche Hall, which was opened in 1973. The building is dedicated to Admiral Russell Randolph Waesche, an eminent Commandant of the Coast Guard during World War II. Modern in style and attractively furnished, the library provides ample reading room and study space for cadets and faculty. Its collections number close to 150,000 volumes, 560 current periodical subscriptions, and more than 40,000 units of microfiche and microfilm. In addition an Audiovisual Study Center, with a collection of nearly 1,800 video cassettes, recordings, and other media, provides a comfortable environment for the use of course-related and recreational AV materials. A partial depository for US government publications, the library maintains a separate collection of federal documents and materials.

Mention should also be made of the Coast Guard Academy's beautiful training ship, the *Eagle*. One of three sister ships built in Germany for the training of German naval officers, it is an 1,800-ton, three-masted auxiliary bark that was acquired as part of the war reparations of 1946. She was originally christened the *Horst Wessel* in 1930 in honor of a Nazi leader killed in a street riot during Adolf Hitler's early rise to power. She operated as a school ship until the early part of World War II, when she was used to carry supplies and passengers between East Prussian ports and Germany proper. It is said that she shot down two Russian aircraft during those short voyages. In January 1946 officers and men of the Academy were sent to Bremerhaven to arrange the details of taking over the ship. Rechristened the *Eagle*, she departed Germany in June 1946, arriving at the Academy in early July.

The *Eagle* is built of German steel on the transverse framing system. Since the fully welded technique had not been developed when she was built, her seams are riveted and the butts welded. She is a beautiful sight under full sail. She has clean-cut lines and a speed reportedly up to seventeen knots. She does not have the lofty masts required of fast sailing vessels, because the Germans wished her to endure and therefore held the stresses down. Besides being a good sailer, she is a fully modern vessel with a diesel engine, electricity, evaporators, air conditioning, the latest electronic devices for navigation and operations, and all the conveniences of modern ships. All cadets attending the Academy have an opportunity to sail on the *Eagle*.

The Coast Guard Academy has excellent facilities for academics, athletics, research, recreation, and day-to-day living. A visit to the Academy is a thrilling experience. You can arrange for a visit during the academic terms. The Coast Guard also has a videotape called "The Will to Succeed." You can rent a copy for a nominal fee by calling College Video Guide, tollfree 800-225-2977.

Entrance Criteria and Application Procedures

The United States Coast Guard Academy is the only one of the service academies that tenders appointments solely on the basis of an annual nationwide competition. No congressional nominations are required, nor are there any geographical quotas. The competition for appointment as a cadet is based on high school rank, performance on either the College Board Scholastic Aptitude Test (SAT) or the American College Testing Assessment (ACT), and leadership potential as demonstrated by participation in high school extracurricular activities, community affairs, or part-time employment.

Eligibility Requirements

To be eligible to compete for an appointment, candidates must fulfill the following requirements:

Age: A candidate must have reached the age of 17 but not have reached the age of 22 by July 1 of the year of entrance. Those under 18 are required to furnish the written consent of parent or guardian.

Citizenship: A candidate must be a citizen of the United States at the time of entry into the Academy. (Foreign nationals nominated by mutual agreement between the US and their native country are exempt from this requirement.)

Marital Status: A candidate must be unmarried at the time of appointment and have no legal obligation resulting from a prior marriage. Any candidate who marries, or is found to be married or to have been married before graduation, shall be required to resign. Refusal to resign will result in dismissal.

Character: A candidate must satisfy the Commandant of the Coast Guard as to moral character and standing in the community and a background that demonstrates positive evidence of responsibility, trustworthiness, and emotional stability. No person who has been dismissed or compelled to resign from any other service academy for improper conduct is eligible for appointment. No person whose discharge from any branch of the military service was under conditions other than honorable is eligible for appointment as a cadet. Substance abuse or homosexual behavior, as in all of the services, will bar you from admittance to the Academy.

Height and Weight: Candidates must be between 60″ and 78″ in height, with weight suitable to physique. They must also successfully undergo a Service Academy Medical Examination. No waivers of educational or physical requirements are granted to applicants.

Scholastic Requirements: A candidate must be a graduate of an accredited high school or preparatory school or be in actual attendance in senior year and have already completed three years' work at such a school. A candidate indicating prospective graduation from a preparatory school or high school must, as a condition of admission, satisfactorily complete the course of study no later than June 30. With the exception of courses completed by correspondence, for which credit has been granted by an accredited school, correspondence schools do not meet the requirements for "Accredited Schools." Certificates issued by correspondence schools will not be accepted. A total of 15 units obtained in high school, preparatory school, or college must be submitted.

You must submit both high school and college transcripts. Because of the great variation in academic standards and credit requirements among schools, the Commandant reserves the right to evaluate each academic record on its individual merit. In general, college credits and high school credits from accredited institutions will be given the same weight for the same amount of work, and in no case will one semester of college work be considered equivalent to more than one unit of high school work.

The Commandant reserves the right to reject any applicant whose assigned grades create doubt as to his or her ability to successfully pursue the Academy course of instruction.

Physical Aptitude: You must, by the last week of Swab Summer and semiannually thereafter, qualify in physical aptitude as determined by a two-part examination. This examination measures neuromuscular coordination, muscular power, speed and strength, cardiovascular endurance, and flexibility. The examination consists of a number of tests such as pull-ups, sit-ups in a two-minute period, standing long jump, 300-yard shuttle run and 1.5-mile run. You should set up a regular program of physical activities before arriving at the Academy

Required courses: The subjects listed below, comprising 6 units, are mandatory and are required for eligibility:

Course	Units
Mathematics (1), algebra to quadratics	1
Mathematics (2), quadratics and beyond	1
Mathematics (3), plane geometry	1
English 1,2,3	3
Total	6

Optional: Further evidence of adequate preparation, amounting to 9 units of optional subjects is required and may be offered from the following groups:

Mathematics
English 4
Social Studies (history, civics, sociology, citizenship, government)
Biological Science (biology, botany, physiology)
Physical Science (physics, chemistry, general science, geology, astronomy)
Foreign Language

A total of not more than 2 units credit will be accepted from any or all of the following groups:

Agriculture	Industrial Problems
Commercial Arithmetic	Journalism
Commercial History	Manual Training
Commercial Law	Mechanical Drawing
Driver Education	Music
Drama	Occupations
Fine Arts	Public Speaking

While not required, Solid Geometry, Trigonometry, Physics, Chemistry, Typing, and a fourth year of English are strongly recommended courses.

and maintain your desired weight. Running, swimming, bicycling, and participating in team sports can help in preparing for the physical demands at the Academy.

Application Procedures

The first step in applying for appointment as a cadet, United States Coast Guard, is to obtain an application for participation in the appointment competition. This form, Application for Appointment as Cadet, can be obtained by writing to Director of Admissions, US Coast Guard Academy, New London, CT 06320. When completed, it should be mailed back to the Director of Admissions. It must be postmarked no later than December 15 to receive consideration for the class entering the following June. Upon receipt of the application form, the Admissions Office will send you the following forms, which must be completed and returned prior to January 15: (a) request for transcript (gray); (b) school officials evaluations (blue); (c) coach evaluation (red); and (d) candidate record and essay (pink, purple).

All candidates for the Academy must take either the Scholastic Aptitude Test (SAT) or American College Testing Assessment (ACT). To register for either test, see your school counselor or write to: Registration Department, College Boards Admissions Testing Program (SAT), Box 592, Princeton, NJ 08540; or Registration Department, American College Testing Program (ACT), Box 414, Iowa City, IA 52243.

You must bear the expenses associated with taking the exams. The examination must be taken by December of the year you are applying. The Coast Guard Academy should be named as one of the colleges to receive the scores. The code number for the Academy to receive the SAT is 5807; the code number for ACT is 0600. It is extremely important that you register for these tests before the deadline established by the testing agency for the December test administration; scores from later tests will not be accepted.

Appointments are tendered on the basis of an annual nation-

A cruise on the Eagle is part of the program at the Coast Guard Academy.

wide competition. All applicants, both civilians and members of the armed forces, participate on an equal basis. Each year the Academy receives approximately 5,000 applications from men and women around the nation. The size of the group that finally enters the academy is approximately 290, making the Coast Guard one of the most selective colleges in the country. However, don't be discouraged by the numbers; only about 3,000 students complete all the application forms. Of those that meet the minimum requirements, about 500 are offered appointments. About 40 percent decline and go to other colleges, universities, or service academies. Therefore, if you meet minimum requirements and complete all the forms, your chances of receiving an appointment are about one in five.

The annual competition is designed to select on a fair competitive basis those candidates who are best qualified and most likely to succeed as cadets and officers in the US Coast Guard. Achievement of those goals depends on: (a) adequate educational background; (b) the possession of aptitudes related to both technical and cultural studies; (c) a sincere interest in the Coast Guard as a career; and (d) relevant personality and physical characteristics.

In addition to the essential virtues of honesty, dependability, and perseverance, the Academy evaluation personnel are looking for evidence of physical stamina, coordination, physical and mental courage, self-confidence, emotional stability, alertness, leadership, and the ability to live and work harmoniously in close contact with others.

The Coast Guard Academy reports to the candidate only the weight assigned each test, the total weighted score achieved on the academic portion of the competition, and the qualifying score for further consideration for appointment.

The criteria used in the selection are the SAT or ACT examination score and the transcripts, evaluations, and questionnaires furnished by each candidate. The results of the academic portion of the competition, consisting of high school class rank and SAT or ACT score, carry a weight of 60 percent in the selection process. The score assigned by the Cadet

Candidate Evaluation Board carries a weight of 40 percent.

An applicant must obtain a minimum score on the SAT or ACT examination. The minimum SAT scores are 500 in math and a total combined score (math and verbal) of 950. The minimum ACT scores are 21 math and total combined score (math and English) of 40.

Cadet Candidate Evaluation Board

The Superintendent of the Academy designates a board of Coast Guard officers charged with the duty of assigning an evaluation mark to each candidate who has satisfied the minimum requirements of the academic portion of the competition. The evaluation includes all the factors known to influence success as a cadet and as an officer. The marks are based on the relative merit of candidates as shown by tests and questionnaires supplied to the Board. The Board's decisions are based on objective information such as the following: (a) the candidate's attitude toward assigned tasks and willingness to work as shown by the consistency and pattern of previous school work; (b) the candidate's previous extracurricular and athletic interests and experience, with particular attention to evidence of leadership and teamwork; (c) the candidate's personal qualities as shown by the reference questionnaire, evaluations and comments by high school counselor, principal, teachers, and similar officials; and (d) the candidate's score on tests of emotional stability, social adjustment, vocational interests, study habits, background, and personality characteristics as may be administered for the purpose. The Board's judgment is final and subject to review only by order of the Commandant. Candidates are offered appointments in the order of their final marks until the vacancies have been filled. A candidate who fails to receive an appointment may compete in subsequent years without prejudice, provided the age and physical qualifications are met. The number of appointments tendered each year is determined by the Commandant of the Coast Guard and is based on the needs of the service. No waivers of educational or physical requirements are granted to applicants.

Medical Examination

A Service Academy Medical Examination is required before a candidate may receive an appointment. Medical examinations are authorized by the Academy as candidates' records become complete. The examination is scheduled and reviewed by the Department of Defense Medical Examination Review Board (DODMERB). Once taken, the examination may be used for other service academies and for four-year ROTC scholarship programs.

Before the formal physical examination, applicants are required to submit a Standard Form 93, Report of Medical History, furnishing a true account of all injuries, illnesses, operations, and treatments since birth. False statements or willful omissions in the form may result in the separation of the candidate from the service on arrival at the Academy or later in the service career.

Formal physical examinations prior to acceptance of candidates must be performed by a US Public Health Service, Navy, Army, or Air Force officer or a military-contracted civilian doctor. Should there be any questions regarding causes for disqualification, information may be obtained by writing to the Department of Defense Review Board, US Air Force Academy, Colorado Springs, CO 80840, Attn: USCG Representative. Candidates must pass a second medical examination during the first week at the Academy, and medically disqualifying defects are cause for disenrollment.

Academic and Military Programs

The Academic Division consists of several departments under the direction and supervision of the Dean of Academics: Engineering, Science, Computer Science, Mathematics, Economics and Management, and Humanities. The Professional

Development Department teaches Nautical Science under the direction of the Commandant of Cadets. These departments are staffed by Coast Guard officers and civilian faculty members. The faculty of the Academy is unique. Collectively, it not only possesses the skills of the traditional college faculty, but also possesses the special skills needed for the training of cadets in the military and seagoing aspects for their careers as Coast Guard officers. The courses presented by these experienced educators assure Academy graduates of a broad foundation in engineering, physical science, liberal arts, and social studies and in those professional studies that specifically prepare graduates for their careers in the Coast Guard.

The Academy recognizes that the majority of graduates must have an engineering or scientific background to meet the needs of the Coast Guard. Present goals are to graduate at least 75 percent in the technical majors: engineering, science, and mathematics. The academic program currently consists of the following majors:

Civil Engineering
Electrical Engineering
Marine Engineering
Marine Science
Mathematical/Computer Science
Government
Management

During the first summer all cadets are assigned to an academic adviser in their major interest area. On occasion, enrollment in a specific major may be limited. In such case selection of students desiring admittance into a major is made by the Dean of Academics, based largely on the student's academic performance in the past. Throughout the four-year academic program the majors normally require a cadet to pass a core program of 26 courses in addition to those required of a major. As cadets progress, they may select additional electives that allow them to pursue further individual academic interests.

Each major provides a sound undergraduate education in a field of interest to the Coast Guard and prepares the cadet to assume initial duties as a junior officer.

Upon graduation, the cadet is awarded the degree of bachelor of science and, if physically qualified, is commissioned by the President of the United States as an ensign in the Coast Guard. After spending a few years of active duty as commissioned officers, postgraduate training and education are available in a wide variety of fields based upon the needs of the service.

The methods of instruction at the Academy are quite varied. Usually classes for core courses consist of 18–22 students; major required courses have as few as 12. Occasionally, the members of an entire year group meet together for general lectures. To supplement its classroom instruction, the Academy faculty makes considerable use of audiovisual aids, laboratory exercises, a computer center, and library resources. The normal periods of instruction are 50 to 75 minutes, with lab periods of two to three hours depending on the subject. On Monday, Wednesday, and Friday, there are four 50-minute morning periods and three 50-minute afternoon periods. The remainder of the morning is used for 50-minute exam periods or military drill. Tuesdays and Thursdays are devoted primarily to exams and laboratory time, with 75-minute periods.

The Academy has an extensive program of individual assistance and counseling to help each student achieve academic success. Each cadet is assigned an academic adviser in his or her major to assist in laying out the academic program and to advise the student as necessary. Additional instruction is available whenever desired. Academic assistance and guidance on an individual basis are provided by instructors and other staff members. Counseling is available for cadets to assist in understanding personal, vocational, educational, and spiritual concerns. Counseling, which is completely confidential, is provided by professional counselors and by the Academy's two chaplains. Pyschotherapy and counseling are also available through the Academy Health Service Division.

Graduation Requirements

To satisfy the requirements for the bachelor of science degree, a cadet must:

1. Satisfactorily complete all core courses.
2. Complete a minimum of 126 credit hours, excluding physical education courses.
3. Attain a cumulative grade point average of 2.0 or better.
4. Be in residence at the Academy for at least four academic years.
5. Maintain a high sense of integrity.
6. Meet the minimum swimming standards.

The tables on the following page show the common first-year curriculum subjects as well as the required core courses for each cadet no matter what the major. In addition to the core courses, cadets must complete a certain number of courses in their chosen major.

Military Program

When you enter the Academy, you become a member of the Corps of Cadets, which gives you the opportunity to develop leadership qualities in preparation for becoming a commissioned officer. Under the supervision of company officers, the Corps of Cadets is self-governing, with cadets having responsibility and authority. The Regimental Commander, under the direction of the Commandant of Cadets, administers the daily routine of inspections, formations, watches, military appearance, and performance of the Corps of Cadets. Unlike students at colleges, Coast Guard cadets must participate in a program of military training and discipline administered by upperclass cadets and officers. Prompt obedience to orders is required; cadets are subject to the Uniform Code of Military Justice at all times.

Cadets receive military and professional training throughout

CORE PROGRAM OF REQUIRED COURSES

The following courses or their authorized substitutes must be completed successfully (or validated) in order to receive a degree:

1113	Intro to Engineering Design	5262	Physics I
1114	Applied Engineering Design	5266	Physics II
1320	Intro to Electrical Engineering	5330	Oceanography
1461	Basic Naval Architechture	6112	Coastal Navigation
2111	English Comp and Speech	6114	Celestial Navigation
2123	Introduction to Literature	6345	Legal Systems
2241	History of the U.S.	6416	The Deck Watch Officer
2263	American Government	6418	The Division Officer
2281	General Psychology	6447	Maritime Law Enforcement
3111	Calculus I	7111	Intro to Computing
3117	Calculus II	7113	Programming Fundamentals
3213	Probability & Statistics	8245	Organizational Behavior
5102	Chemistry I	8311	Economics
5106	Chemistry II		

Authorized substitutions for core courses

The following substitutions for core courses are authorized for any major:

Course		Substitution	
1320	Intro to Electrical Engr	1220	Electrical Engineering
1461	Basic Naval Architecture	1342	Principles of Naval Architecture
8311	Economics	8215	Macroeconomic Principles

1. Other substitutions can be made only when the substitution is specifically authorized or required in a major program. Honors Calculus, Physics, and/or Chemistry may be substituted for the core requirement.
2. Directed studies may be substituted for any major requirement, major area elective, or related area elective with the approval of the major coordinator.
3. Major Area Electives offered for other year groups in the same major may also be used to meet the Major Area Elective requirements for this year group. However, the same course cannot be used in both the Major Requirement and Major Area Elective categories.

FIRST YEAR PROGRAM (generally the same for all cadets)

Fall Semester		Spring Semester	
1113 (7111)	Engineering Design **OR** Intro to Computing	2123	Introduction to Literature
2111	English Comp and Speech	3117	Calculus II
3111	Calculus I	5106	Chemistry II
5102	Chemistry I	6114	Celestial Navigation
6112	Coastal Navigation	7111 (1113)	Intro to Computing **OR** Engineering Design
	Physical Education		Physical Education

their four years at the Academy, including four summer periods of training. During the first summer as Fourth Classmen, known as Swab Summer, the emphasis is on indoctrination. In a sense, it is similar to boot camp. Cadets take courses in all phases of military activity and are required to pass physical fitness and swimming tests. If you are not in shape when you arrive for Swab Summer, this seven-week period will be extremely difficult and stressful. The highlight of this summer is a one-week cruise aboard the *Eagle*.

As a Third-Class (3/c) cadet, your summer will be spent at sea aboard *Eagle* for five weeks and at a variety of one-week programs that give you hands-on training in sailing, seamanship, navigation, small arms handling, damage control, and fire fighting. Use of the Academy's numerous small craft, including sailboats, small power boats and 65-foot tugboats is maximized in this training. You may have the opportunity to travel to many American, Caribbean, European, or other ports around the world.

During either your 3/c or 2/c summer, you will spend several weeks at an operational Coast Guard Search and Rescue Station, a Marine Safety Office, an Environmental Response Strike Team, or on a Coast Guard Patrol Boat and will participate in two weeks of Coast Guard aviation orientation. Cadets study basic theory of flight, operational employment of aircraft, and search and rescue planning. You will have an opportunity to handle the controls, navigate aircraft, visit the "dunker," receive water survival training, and experience the problems faced by pilots firsthand.

As a Second Class cadet, you will sail *Eagle* for five weeks in a senior petty officer role, spend two weeks sailing on the Academy's 44-foot Luder yawls, attend a one-week Leadership School, and help in the training of the incoming 4/c cadets. You may also participate in one of the Boys or Girls State programs throughout the country.

During the First Class summer program, you will spend ten weeks aboard a Coast Guard cutter where your role will be as a junior officer afloat. This participation in actual Coast Guard operations will help prepare you for your duties after graduation.

Military training is an integral part of your four years at the Academy. You are immediately introduced to the Honor Concept, to barracks life, and to living and performing in a highly regulated and strenuous environment. Formal training in leadership begins with a course on human behavior. In addition to this course, intense instruction in leadership is given to all cadets. Some cadets are chosen to participate in the summer cadre program, during which they serve as leader and teacher for the entering class of cadets and have the opportunity to put into practice those skills and concepts learned in class. Before commissioning, cadets take a course in organizational behavior that stresses the role of the individual, group, organization, and leadership in the total management process. All cadets also take a course entitled Legal Systems, which stresses the fundamental nature of law and essential professional training in military justice and nonjudicial punishment. These two courses round out the formal classroom training to develop leadership skills.

In addition to formal classroom work, leadership programs are conducted by the commissioned officers for the cadets. During these programs, case studies are discussed with officers who have worked in the Coast Guard and have firsthand knowledge of the challenges and responsibilities to come. The summer training cruises provide extensive practical experience to cadets as they work in roles of petty officers and officers. The entire four years are spent in a military environment—one of accountability for one's actions as well as for the cadets under one's command. The Academy experience is designed to instill the traits and provide the skills needed to become a successful Coast Guard officer.

Cadet Life

Life at the Coast Guard Academy, like life at all the service academies, is highly structured and busy. It always seems that you do not have enough time to do what is required of you. You must remember that you are a member of a military organization, and the nature of any military organization requires that each individual and each unit be responsive to orders from above. This is known as the chain of command, and following the chain of command is essential to the smooth efficiency of all organizations. Underlying the overall concept of military discipline is the Honor Concept. It is a basic vital force in cadet life; cadets at the Academy are men and women of honor who neither lie, cheat, steal nor attempt to deceive. Training in the Honor Concept begins during Swab Summer and continues throughout the four years. Violations of the concept often result in severe punishment or disenrollment, which reinforces the importance of integrity. Through the

Men and women midshipmen parade together at the Coast Guard Academy.

Honor Concept, cadets are expected to develop trust among themselves and their coworkers.

Daily life at the Academy is full. The typical day begins at the crack of dawn, 6 a.m., with reveille. You prepare for daily room inspection and breakfast. Morning classes begin at 7:55 and continue until 11:45. Personnel inspection is held at the noon formation. Lunch is next, followed by afternoon classes from 12:50 to 3:35 p.m. The hours between 3:35 and 6:45 are spent in intramural or varsity sports, extracurricular activities, and extra academic instruction for cadets who need additional help. Dinner is 6:45. Time between 7:00 and 8:00 is set aside for personal use, during which cadets may study or attend to personal matters. Study time is between 7:00 and 10:00. Taps is at 10:00, but cadets may continue to study until midnight. Saturday mornings are spent in professional training and indoctrination. During the fall and spring months, formal regimental parades and inspections are held on Friday afternoons.

Liberty, which means permission to leave the Academy grounds, is granted on Saturday afternoon and evening, and again on Sunday. First Class (senior) cadets are given liberty on Wednesday afternoons at 4:00 and all cadets except Fourth Class (freshman) cadets are granted liberty on Friday afternoons at 4:00. Civilian clothes are authorized on liberty for First and Second Class cadets. Cars may be maintained at the Academy by First Class cadets only.

All cadets live on campus at the Academy. You share a room with another cadet and you are responsible for its cleanliness and neatness. During the year, a cadet is usually granted six weeks of leave and long weekends that may be spent away from the Academy if he or she has no other military obligation. Provision is made for religious services of all faiths. Catholic and Protestant chaplains provide the opportunity for Christian worship and service. Jewish cadets and those from other religious groups are assisted in finding religious resources in the civilian community. Attendance at religious services is voluntary.

Following is a general outline of the schedule of activities at the Academy over a four-year period:

SCHEDULE OF ACTIVITIES

FIRST YEAR (Fourth Class)

Summer Program - Entrance Orientation, and Short Cruise - 7 weeks
Fall Academic Semester - 15 weeks
Examinations - 1½ weeks
Christmas Leave - 2 weeks
Spring Academic Semester - 15 weeks
Spring Leave - 1 week
Examinations - 1½ weeks
Graduation Activities - 1 week

SECOND YEAR (Third Class)

Summer Program - Long Cruise - 11 weeks
Leave - 3 weeks
Fall Academic Semester - 15 weeks
Examinations - 1½ weeks
Christmas Leave - 2 weeks
Spring Academic Semester - 15 weeks
Spring Leave - 1 week
Examinations - 1½ weeks
Graduation Activities - 1 week

THIRD YEAR (Second Class)

Summer Program - Training at Shore installations - 11 weeks
Leave - 3 weeks
Fall Academic Semester - 15 weeks
Examinations - 1½ weeks
Christmas Leave - 2 weeks
Spring Academic Semester - 15 weeks
Spring Leave - 1 week
Examinations - 1½ weeks
Graduation Activities - 1 week

FOURTH YEAR (First Class)

Summer Program - Long Cruise and Surface Operations Program - 11 weeks
Leave - 3 weeks
Fall Academic Semester - 15 weeks
Examinations - 1½ weeks
Christmas Leave - 2 weeks
Spring Academic Semester - 15 weeks
Spring Leave - 1 week
Examinations - 1½ weeks
Graduation Activities - 1 week

Physical education is an important part of life at the Academy. All cadets are required to pass an entrance physical fitness test and demonstrate minimum swimming skills. Thereafter, cadets must pass a standard physical fitness test semiannually. Each cadet is required to take physical education courses each semester. Courses are offered in developmental, recreational, and professionally oriented subjects. Courses offered include:

Survival at Sea
Advanced Swimming
Advanced Lifesaving
Physiology of Fitness
Fundamentals of Conditioning
Gymnastics
Personal Defense I
Personal Defense II
First Aid/CPR
Golf
Tennis
Racquetball

Cadets are also required to participate in either intramural or intercollegiate athletics in two out of three sporting seasons. For men, there is a full slate of intercollegiate sports from which to choose, including football, sailing, soccer, cross-country, tennis, basketball, wrestling, swimming, track, golf, and baseball. Women cadets are offered intercollegiate volleyball, basketball, softball, gymnastics, track, and other sports.

A large number of extracurricular activities are available. Formal and informal dances, parties, movies, and other social functions are scheduled regularly. A strong musical activities program includes the Cadet Protestant and Catholic Choirs, the Coast Guard Academy Idlers and Icebreakers, male and female singing groups respectively, the Windjammers Drum and Bugle Corps, the Cadet Jazz Ensemble, and several small groups playing contemporary music. A cadet who plays a musical instrument may bring it to the Academy. The Hockey Club, Ski Club, Speech and Debate Club, Outdoor Club, Political Affairs Association, Racquetball Club, and Officers' Christian Fellowship are typical of over twenty extracurricular organizations active at the Academy. *Howling Gale*, the cadet magazine, and *Tide Rips*, the yearbook, offer experience and pleasure to cadets with interests in writing.

While at the Academy, a cadet receives approximately $525 per month. The pay, although taxable, is not a wage or salary in the usual sense. Tuition is paid by the government. The money furnished by the government to each cadet is used for uniforms, equipment, textbooks, and other expenses incidental to training and education. If you are appointed and decide to accept, a $1,500 entrance fee is required to defray initial costs. This fee can be deferred by the Director of Admissions in hardship cases. Cash pay and allowances during the four years are disbursed and expended as directed by the Superintendent. Any funds remaining in a cadet's account are given to the cadet upon graduation. Cadets receive the following amounts each month in a checking account for personal use: 4/c—$80, 3/c—$150, 2/c—$210, 1/c—$260.

You should find out as much about the Academy as you can before making decisions about applying. You should also find out as much as you can about what kind of jobs you will receive when you graduate from the Academy. Visit the Academy if possible. Each Friday afternoon at 1:15 p.m. Admissions sponsors a presentation that includes a film, a question-and-answer session, and a tour of the grounds. Appointments for the program may be obtained by writing to the Director of Admissions or by calling (203) 444-8503 between 8 a.m. and 4 p.m. EST. Talk to recent graduates or active duty Coast Guard personnel about the Academy. A videotape called "The Will to Succeed" is available; you can rent a copy by calling 800-225-2977. You will increase your chances for success by finding out as much as possible about the Academy. It is an extremely tough program; at least two out of five cadets who start the program do not complete the four years and graduate. You must have the will to succeed. For additional information, write to Director of Admissions, US Coast Guard Academy, New London, CT 06320, phone (203) 444-8501.

Chapter VII

The United States Merchant Marine Academy

History

The United States Merchant Marine Academy is one of the five federal service academies. It is sometimes overlooked when considering service academies, but it is a national institution, military in character, and an accredited, four-year, degree-granting college. In June 1936, Congress enacted the Merchant Marine Act, which formally declared US government interest and involvement in merchant marine operations, including the formation of a merchant fleet "manned with trained and efficient citizen personnel." In March 1938 the US Maritime Commission issued a General Order assuming responsibility for federal maritime cadet training and creating a US Merchant Marine Cadet Corps. Training was first given aboard merchant ships and later at temporary shore establishments pending the acquisition of permanent facilities. The Walter P. Chrysler estate at Kings Point, New York, was selected as the site for the Academy in March 1942, and construction was begun the following May. Fifteen months later the work was virtually completed, and the United States Merchant Marine Academy was dedicated on September 30, 1943.

World War II required the Academy to forgo normal operation and devote all its resources to meeting the emergency personnel needs of the merchant marine. The enrollment was increased to 2,700, and the course of instruction was shortened from four years to two years. Notwithstanding the war, shipboard training continued to be an integral part of the curriculum, and midshipmen served at sea in combat zones the world over. Two hundred twelve midshipmen and graduates gave their lives in service to the country, and many others survived torpedoings and bombings. Seven midshipmen and one graduate were awarded the Merchant Marine Distinguished Service Medal, one of the nation's highest decorations for conspicuous gallantry and devotion to duty. By the end of the war the Academy had graduated 6,634 officers. In the closing days of World War II plans were formulated to establish a four-year, college-level program to meet the peacetime needs of the merchant marine. At the end of the war, in August 1945, the four-year course was immediately instituted with the September class of midshipmen.

The Academy has since grown in stature and has become one of the world's foremost institutions in the field of maritime education. Authorization for awarding graduates the bachelor of science degree was granted by Congress on August 18, 1949. The Academy was accredited as a degree-granting institution by the Middle States Association of Colleges and Schools on November 26, 1949, and was accredited by the Engineering Board of Engineering Technology. The Academy was made a permanent institution by an Act of Congress on February 20, 1956, and its operation was placed under the authority of the Department of Commerce; it is now under the Department of Transportation. On July 16, 1974, the USMMA became the first federal academy to admit female students. Fifteen women reported for that class. The other federal academies did not admit women until almost two years later. In October 1980 the Maritime Education and Training Act was passed by Congress, establishing for the first time a legal obligation for merchant marine and US Naval Reserve service by Academy graduates. Today, Kings Point graduates are serving with distinction in all sectors of the maritime industry—as ship's officers, steamship company executives, admiralty lawyers, marine underwriters, shipbuilders and repairers, naval architects and oceanographers, and career officers in the United States Armed Forces.

The reasons for the establishment of a Merchant Marine Academy almost fifty years ago are just as valid today as they were then. As a result of advanced technology in ship design, modern management techniques, reliance of all nations on balanced international trade, and growing interest in the oceans as a source of natural resources and food, the challenge of the sea is greater than ever.

A competitive merchant fleet of highly productive ships guarantees our nation access to foreign sources of raw materials and to foreign markets for our manufactured goods. In addition, such a fleet is vital for national security; in time of war or national emergency, merchant ships are our "fourth arm of defense" and carry the brunt of delivering military supplies across the oceans to our forces and allies. The United States currently imports some 85 percent of seventy-seven strategic and critical commodities required to meet its economic and defense needs. Although the United States has less than 6 percent of the world's population, it channels more than one third of the world's raw material output to the American economy. Ninety-nine percent of these materials are transported by ship.

The major component of our maritime industry is the United States Merchant Marine, the fleet of privately owned, American-flag vessels that transport cargo to the major ports of the world. But a strong maritime industry consists of more than seagoing ships. It also includes large numbers of tugs, barges and river craft, steamship companies, ports and terminals, shipyards, marine insurance underwriters, ship chartering firms, admiralty lawyers, and a vast array of other specialized firms directly or indirectly engaged in maritime-related commerce, engineering, research, and management.

More than anything else, a productive and competitive merchant fleet and a strong maritime industry require leadership—men and women who are intelligent, ambitious, well educated, and competent. The prime raison d'être for the

Merchant Marine Academy is to ensure that such officers are available to meet the challenges of the present and the future.

Mission

As we have seen, the mission of the US Merchant Marine Academy had its origins in the Merchant Marine Act of 1936, which recognized that an American merchant marine is essential for the nation's economic well-being and security. The act states that US merchant ships must be crewed by professional, qualified citizens. Congressional actions over the years have reaffirmed the importance of the merchant marine to the nation's security. The mission of the Merchant Marine Academy is:

- To serve the economic and national security interests of the United States and to foster a strong, competitive and safe American merchant marine through nationwide recruitment, training, and graduation of outstanding young Americans with definite ambitions to serve as merchant marine officers and Naval Reserve officers;
- To be a prime source of leadership for maritime America;
- To provide its midshipmen with both the academic and practical shipboard training to prepare them, upon graduation, for immediate employment as watch-standing third mates and/or third assistant engineers on US-flag merchant vessels, and to be naval officers commissioned at graduation as Ensigns in the US Naval Reserve;
- To give them the sound education in theory underlying the skills required of a ship's officers so that they may qualify for positions of greater responsibility aboard ship and may be able to utilize and contribute to the rapidly changing management technology and transportation systems in the maritime field;
- To offer them the broadest possible program of general education consistent with the professional character of the Academy's mission.

- To provide them with a sound background in management and technical skills so that, after serving at sea, they can be promoted to leadership positions ashore where they can further guide and direct the development of our nation's merchant marine, maritime industry, and intermodal transportation systems;
- To develop in them the qualities of self-discipline, responsibility, and leadership for effective citizenship and successful maritime careers in both peace and war;
- To set an international standard of excellence in maritime training that may be made available to citizens of other nations, in limited numbers, when deemed in the interests of the United States.

It is a very ambitious mission; the Merchant Marine Academy is proud of what it has accomplished in a short period of time.

Facilities

The Academy occupies eighty acres of land at Kings Point, on the north shore of Long Island, about twenty miles east of New York City. The campus and facilities were planned for a normal enrollment of 1,000 midshipmen (the term is applicable to women as well as men). The design of the buildings is simple yet functional, and the campus has been laid out to take full advantage of the picturesque landscape of the north shore. The buildings and walks are named for persons whose deeds brought fame to the merchant marine or to the Academy.

On the slope looking toward Long Island Sound stands a monument erected to the memory of the two hundred twelve midshipmen who lost their lives at sea during World War II. The War Memorial is the focal point of the western part of the campus.

Around the monument, adjoining the Sound, are grouped the following buildings and facilities: indoor and outdoor swimming pools; boat basin and pier facilities; the Wooster Building, housing the Public Works Department; the Fitch building, housing the National Maritime Research center; Gibbs Hall, housing science and engineering laboratories; and Samuels Hall, with classrooms and laboratories for nautical science. A beautiful Memorial Chapel, honoring all men and women of the merchant marine, stands on a grassy knoll to the south of the War Memorial.

Wiley Hall, facing Long Island Sound, is the center of administrative activities and contains the offices of the Superintendent, Deputy Superintendent, Academic Dean, Commandant of Midshipmen, Admissions, and other members of the administrative staff. East of Wiley Hall lies the center of the Academy campus, marked by one of the nation's tallest unguyed flagpoles, 176 feet 6 inches high. Surrounding the flagpole are Fulton Hall, the engineering and science building; Bowditch Hall, housing the Department of Humanities, the Department of Marine Transportation, and the auditorium; and the Schuyler Otis Bland Memorial Library. Delano Hall, the midshipman dining room, and six dormitory buildings—Jones, Barry, Rogers, Cleveland, Murphy, and Palmer Halls—complete the circle of buildings enclosing the main campus. The dormitories and dining room and connected by an underground promenade, which contains the midshipman lounge and canteen, uniform and varsity shop, laundry facilities, bank, barbershop, and ship's service store.

On the perimeter of the Academy are the athletic fields; Furuseth Hall, containing the nuclear engineering facilities, the Department of Naval Science, Shipboard Training, and the Administration Department; O'Hara Hall, which has a spacious gymnasium, an indoor swimming pool, and athletic facilities; the Patten Infirmary; and Land Hall, the midshipman recreation and activities building. Other buildings on the grounds are used as residences by the Superintendent and officers of the administrative staff.

The Schuyler Otis Bland Memorial Library is a three-story building; it has shelves for 160,000 volumes and can accommodate 300 readers. Each floor has a book stack area with adjacent reader areas. Separate rooms are provided for special collections, including current periodicals, fiction, microtexts, charts and atlases, US government publications, archives, and rare books. The library provides microfilm readers, a microfiche reader, a reader-printer, and a photocopier. A program of library-sponsored cultural events and exhibits supplements the provision of educational materials. Study facilities include carrels, seminar rooms, small group discussion rooms, faculty studies, and a typing room. Also in the library is a rotating 75-inch geophysical globe that treats ocean areas in full detail.

Two other facilities of note at Kings Point are the National Maritime Research Center and the Computer-Aided Operations Research Facility (CAORF). The National Maritime Research Center, under the direction of the Office of the Senior Advisor for Research and Development of the Maritime Administration, was established at the Academy in June 1971. It mission is to serve as a national center for the coordination and dissemination of information on new technologies; to serve as a test and evaluation center for products and innovations resulting from research and development programs of the Maritime Administration and the maritime industry; and to provide technical support to the Maritime Administration's research programs.

The Computer-Aided Operations Research Facility, in Samuels Hall, has been in operation for fifteen years. This facility, which contains a ship-handling simulator, can duplicate almost any navigational problem that could be encountered by a real ship, taking into account the human factor. It features a full-scale bridge surrounded by a 240-degree field of vision screen to display a visual scene generated by computers. Sea and weather conditions as well as port and traffic situations are realistically simulated. The CAORF, which is utilized to train midshipmen, provides sophisticated simulation of shipboard maneuvering and operational situations under controlled conditions. It performs research to improve ship performance and productivity, reduce ship collisions and groundings, improve the training and certification of watch standers, assist in the development of ports and new port operations, define bridge system effectiveness, and develop standardized bridge designs. It is a state-of-the-art training aid.

Entrance Criteria and Application Procedures

All candidates must meet certain requirements of citizenship, age, and moral character, but the Academy considers qualified applicants without regard to race, color, sex, or national or ethnic origin.

Citizenship: Candidates must be citizens of the United States, either by birth or by naturalization, except for foreign candidates.

Age: Candidates must be at least 17 and not have passed their 25th birthday on July 1 of the year of admission.

Moral Character: Candidates must be of good moral character with no record of substance abuse, homosexual behavior, etc.

Naval Reserve Midshipman Requirements: Candidates must meet the physical, security, and character requirements for appointment as Midshipman, USNR.

Scholastic Requirements: Candidates must have graduated from an accredited secondary school or the equivalent and must present at least 15 units of credit. Seven units are required as follows:

3 units of English
3 units of mathematics (algebra, geometry, trigonometry)
1 unit of physics or chemistry with a laboratory

These quantitative requirements are minimal and should serve as a basis upon which to build a strong high school program. It is recommended that all candidates take four years of mathematics and both physics and chemistry. Courses in mechanical drawing and machine shop are helpful, as well as typing and computer courses.

By March 1 of the year in which you are seeking admission, you must submit evidence with your official application show-

Picturesque Kings Point is the home of the U.S. Merchant Marine Academy.

ing completion or scheduled completion of academic requirements. All required courses must be completed by June 15 or by the date of graduation from high school. No extensions in time will be granted.

Candidates for admission to the Academy must be nominated by a member of Congress or another nominating authority. Unlike the other service academies that require nominations, the Merchant Marine Academy does not accept Presidential or Vice Presidential nominations. Candidates can be nominated only by nominating authorities from their state or geographical area. Residents of the Northern Mariana Islands and American Samoa are nominated by the Governor. The representative to the House of Representatives for Guam, the Virgin Islands, the District of Columbia, the Commonwealth of Puerto Rico, and American Samoa, and the Panama Canal Commission for US citizens residing in Panama, may each annually nominate ten candidates for the Academy. Special legislation also permits the appointment of midshipmen from the Latin American republics, the Trust Territories of the Pacific, and nations other than the United States. Additionally, up to 30 appointments can be made to candidates from nations other than the US on a reimbursable basis to the Maritime Administration. These foreign student candidates must be approved by the Department of State and be sponsored or approved by their nation. Total enrollment at the Merchant Marine Academy is approximately 900 midshipmen, with about 300 new candidates entering each year.

Nominating officials select nominees by any method they wish, including a screening examination. The examination may be given as early as July of the year before appointment is sought. The test is for nomination only and should not be confused with the examinations required for appointment. Nominees must reside in the state, area, or territory that the nominating official represents. This system insures that a genuine cross-section of American youth will be represented at the Academy at all times.

You should apply early for a nomination. Some nominating

officials establish deadline dates for the receipt of requests in order to allow adequate time for processing and evaluating the applications. The best time for application is May of your junior year in high school. A sample letter to members of Congress is given in the Appendix. Nominating officials submit the names of their nominees to the Academy between August 1 and December 31 of the school year preceding that in which admission to the Academy is sought.

Application to the Academy can be made even before requesting a nomination from a member of Congress. Filing an application early permits more rapid processing of the admission file. An application form, as well as information about the Academy, can be obtained by writing to Admissions Office, United States Merchant Marine Academy, Kings Point, New York 11024–1699.

All candidates are required to take either the College Board's Scholastic Aptitude Test (SAT) or the American College Testing Program's Test (ACT). Testing must be completed by the first test date of the year in which admission is sought, unless permission for a later test date is requested and received in writing from the Director of Admissions. All tests should be taken within sixteen months before the month of admission. It is the candidate's responsibility to register for the exams. Registration instructions are contained in information bulletins available at no cost to secondary schools. Members of the armed forces should find copies available in their unit's education offices. Bulletins may be obtained by writing to:

College Board
PO Box 592
Princeton, NJ 08541
or
PO Box 1025
Berkeley, CA 94701

American College Testing Program
PO Box 168
Iowa City, IA 52240

Candidates must request the testing agency to submit their test scores to the Academy; the College Board code number is 2923, and the ACT code number is 2974.

Physical examinations are conducted by a service academy examining facility designated by the Department of Defense Medical Review Board, and a final decision on a candidate's physical qualifications is made by that Board. Upon reporting to the Academy, all new midshipmen are subject to an eye examination. Those who fail to meet the visual requirements are disenrolled. Failure to meet vision and color standards in the eye examination is the most common disqualifier. No waivers of the general, scholastic, or physical requirements are granted.

Candidates are appointed competitively by the Academy for the vacancies allocated to their state or geographical subdivision. Each state has a quota proportionate to its representation in Congress. After the principal appointees have been selected, the remaining qualified candidates are designated as alternates, to be appointed in order of merit should vacancies occur within their state. In the event there are insufficient candidates within a state's quota, appointments to fill the unmet quotas are made from the national list of alternates, ranked in order of merit.

A candidate's competitive standing is determined by his or her test scores on the SAT or ACT examination, high school class rank, academic record, evidence of leadership potential, interest in a maritime career, and other factors considered to be effective indicators of motivation and probability of successful completion of the training. Bonus points are given to candidates with six months or more of sea service aboard merchant vessels. Because so many applicants apply for a limited number of vacancies, the Academy can be very selective in its appointments. Scores may change each year, but an average candidate has a SAT verbal score of 530 and a mathematics score of 600, ranks in the top 10 percent of his or her class, has good recommendations, and has exhibited leadership characteristics in extracurricular activities in high school.

The Academy is the final authority for awarding appointments. It is possible to receive a nomination for the Academy but not receive an appointment. Candidates are chosen on a "whole-person" evaluation system, on the basis of: (1) SAT or ACT results; (2) overall high school or academic record; (3) class rank; (4) extracurricular activity record; (5) leadership potential; and (6) motivation toward a maritime career, interest in the Academy, industriousness, citizenship, and recommendations from high school counselors, teachers, and/or principal.

Outstanding candidates may be notified of appointment early, after complete evaluation of their qualifications. All other candidates are notified of their status on or about May 1 of the year in which they seek nomination. Principal candidates are invited to visit the Academy in May for a briefing program and an opportunity to meet and discuss the Academy's program with midshipmen, faculty, and administrative personnel. Attendance is voluntary. Travel arrangements and expenses must be borne by the candidate.

Academic and Military Programs

The United States Merchant Marine Academy offers a rigorous four-year program leading to a bachelor of science degree, a US Coast Guard license, and a commission as an Ensign in the US Naval Reserve. The curriculum is demanding, comprehensive, and stimulating. Each graduate is professionally competent, trained for leadership and responsibility, and well rounded intellectually.

All midshipmen must satisfy the core curriculum requirement and must elect a major program. Selection of the major program determines which US Coast Guard license training program will be taken: deck or engineering officer. Finally, each midshipman spends approximately a year at sea.

Core Curriculum

This requirement has several components:

A. Mathematics—Four courses in calculus (16 quarter hours)
B. Science—Four courses in physics and two in chemistry (22 hours)
C. English—Three courses (9 hours)
D. Humanities and History—All midshipmen take three courses in history (9 hours) and one of the humanities sequences noted below.
   - Engineering majors. A three course sequence in humanities or a three course sequence in comparative cultures (9 hours)
   - Marine Transportation majors. A four course sequence in humanities, or a four course sequence in comparative cultures, or a four-quarter sequence in Spanish (12 hours)
   - Dual License majors. A four course sequence in humanities (12 hours)
E. Naval Science—Four courses (12 hours)
F. Physical Education and Ship's Medicine—Eight courses in physical education and ship's medicine (8 hours)
G. Computer Science—Two courses (3.5 hours)

Specific courses meeting these requirements are identified later under departmental headings.

Major Program

Midshipmen must select one of the four major programs.

A. Marine Engineering.
B. Marine Engineering Systems—Accredited by the Accreditation Board for Engineering and Technology.
C. Marine Transportation—A combination program consisting of management and nautical science.
D. Dual License—A combination consisting of marine engineering and marine transportation.

U.S. Coast Guard License Programs

Deck Officer—training in nautical science as preparation for the third mate's license examination. Required of marine transportation majors.

Engineering Officer—training in marine engineering as preparation for the third assistant engineer's license examination. Required of all engineering majors. Note that dual license majors take both license preparation instruction and license examinations.

Sea-Year Training

All midshipmen spend six months at sea during both their second year and third year aboard commercial ships, training for license preparation. This program is explained more fully under Shipboard Training.

Electives

Marine Engineering Systems majors and Dual License majors are not required to take electives, but nevertheless are encouraged to do so since electives are an important means of pursuing intellectual and professional interests. Majors in Marine Transportation and Marine Engineering must complete 18 quarter credit hours of electives. Note that independent study courses are considered electives.

Graduation Requirements

The minimum requirements for graduation are as follows:

- Pass the required resident and sea project courses. A four-year course of study is required by statute.
- Earn the minimum number of quarter credit hours required by the curriculum in which the midshipman is enrolled.

Marine Transportation

D128	Nautical Science II	0.5
D129	Nautical Science III	5
L281	Accounting for Management	3

At the end of the fourth class year (June/July) half of marine transportation and engineering midshipmen go to sea (B-splits). The other half go on leave, and return in July for two more resident quarters (A-splits). In December/January of the third class year, the two groups switch places. This pattern continues in second class year.

Midshipmen in the Dual License curriculum go to sea following completion of their fourth class year (June/July), and return in December/January of their third class year.

They remain at the Academy for the next four consecutive academic quarters, and return to sea in December/January of their second class year. They return to the Academy to complete their resident curriculum as first classmen following the end of their second sea period.

The Shipboard Training Program

As part of their professional training, midshipmen participate in a cooperative educational program consisting of two quarters of the sophomore year and two quarters of the junior year at sea (approximately five months for each sailing period) aboard commercially operated merchant ships. Every effort is made to assign midshipmen to several different vessels during their two periods of training. They become familiar with the performance and operating characteristics of various classes of ships and with the diverse operating requirements of different trade routes, while at the same time gaining valuable practical experience in the performance of shipboard duties.

The shipboard training program provides midshipmen with the opportunity to use a ship as a seagoing laboratory. Midshipmen are given a study guide called a "sea project" and,

Program Sequence
Fourth Class (Freshman) Year

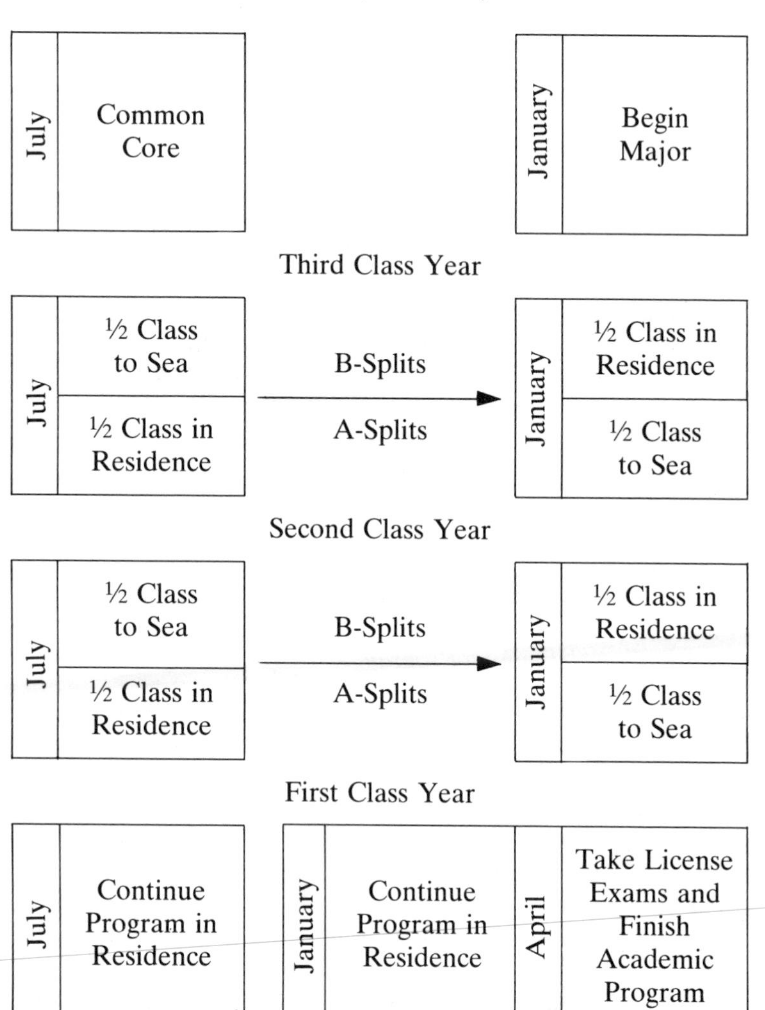

administered by the Department of Naval Science at the Academy. The department is staffed by officers and enlisted personnel who are assigned to the Academy by the Department of the Navy. The Chief of Naval Education and Training prescribes the naval science curriculum and furnishes required textbooks, references and training aids.

The performance of midshipmen is monitored by the head of Department of Naval Science. In the event a midshipman fails to display those qualities of leadership, character, and aptitude of a prospective naval officer, the department head will make a recommendation to the Secretary of the Navy, via the Superintendent of the Academy and the Chief of Naval Education and Training, that the midshipman's appointment be terminated. The midshipman may also be separated from the Academy.

Maritime and Naval Reserve Obligations

Each appointee to the Academy who is a citizen of the United States shall as a condition of appointment sign an agreement:

A. To complete the course of instruction at the Academy, unless the individual is separated by the Academy;
B. To fulfill the requirements for a license as an officer in the merchant marine of the United States on or before the date of graduation from the Academy of such individual;
C. To maintain a license as an officer in the merchant marine of the United States for at least six years following the date of graduation from the Academy of such individual;
D. To apply for, and accept if tendered, an appointment as a commissioned officer and serve in the United States Naval Reserve, the United States Coast Guard Reserve, or any other reserve component of an armed force of the United States, for at least eight (8) years following graduation from the Academy of such individual;

E. To serve the foreign and domestic commerce and national defense of the United States for at least five years following the date of graduation from the Academy:
   1. As a merchant marine officer serving on vessels documented under the laws of the United States or on vessels owned and operated by the United States or by any state or territory of the United States;
   2. As an employee in a United States maritime-related industry, profession or marine science (as determined by the Secretary of Transportation), if the Secretary of Transportation determines that service under *1* above is not available to the individual;
   3. As a commissioned officer on active duty in an armed force of the United States or in the National Oceanic and Atmospheric Administration; or
   4. By combining the services specified in *1*, *2*, and *3* above; and

F. To report to the Secretary of Transportation on the compliance by the individual to these requirements.

Failure to Fulfill Obligation

If the Secretary of Transportation determines that any midshipman who has entered the second class (junior) year or beyond has failed to fulfill the part of the agreement described in paragraph *A* above, he or she may be ordered by the Secretary of the Navy to active duty in an enlisted status to serve for a period not to exceed two years. In cases of hardship, the Secretary of Transportation may waive this requirement.

A midshipman entering the Academy from the regular or reserve component of the Navy will not have his or her enlistment or period of obligated service terminated because of the acceptance of an appointment as a midshipman. If such a midshipman fails to complete the course of instruction at the Academy, this midshipman will resume his or her prior enlisted status and shall complete the period of obligated service. All service as a midshipman is counted as service under that

enlistment or period of obligated service. However, completion or partial completion of a service obligation acquired by prior enlistment in no way exempts a midshipman who has entered his/her second class year or beyond and fails to complete the course of instruction at the Academy from being transferred to the reserve component and ordered to active duty, the same as a midshipman who enters the Academy directly from a civilian status.

If the Secretary of Transportation determines that any individual has failed to fulfill any part of the agreement described in subparagraphs *B*, *C*, *D*, *E*, or *F* above, he or she may be ordered to active duty to serve a period of time not less than three years and not more than the unexpired portion (as determined by the Secretary of Transportation) of the service required by subparagraph *E* above. The Secretary of Transportation, in consultation with the Secretary of Defense, shall determine in which service the individual shall be ordered to active duty to serve such period of time. In case of hardship, the Secretary of Transportation may waive this requirement.

The Secretary of Transportation may defer the service commitment of any individual pursuant to subparagraph *E* for a period of not more than two years if such individual is engaged in a graduate course of study approved by the Secretary of Transportation, except that any deferment of service as a commissioned officer pursuant to subparagraph *E3* must be approved by the secretary of the department that has jurisdiction over such service.

Commission as Ensign, Merchant Marine Reserve, United States Naval Reserve

Upon graduation and Coast Guard licensing as a third mate or third assistant engineer, and with the recommendations of the head of the Department of Naval Science and the Superintendent of the Academy, midshipmen are commissioned as Ensigns, Merchant Marine Reserve, U.S. Naval Reserve (MMR, USNR). This commissioning program, initiated in

1978, recognizes the special qualifications of licensed deck and engineering officers who are members of the Naval Reserve. The aim of this program is to provide an interface between U.S. naval forces and the maritime industry, thereby providing a group of well-trained seagoing professionals who can cooperate with naval forces when needed.

After accepting an appointment as an Ensign in the Merchant Marine Reserve, United States Naval Reserve program, a graduate must perform satisfactorily in the Naval Reserve for eight years. He or she may apply for and, if accepted, serve on fulltime active duty as an officer in the Navy for three consecutive years. Any portion of the eight-year period not served on active duty will be served on inactive duty, during which time the following will apply to every male and female graduate:

1. He or she will be a member of the Ready Reserve (USNR-R).
2. He or she will sail on a U.S. Coast Guard-issued license at sea for at least four months during every two complete consecutive years following the acceptance of the commission unless such requirement is waived by the Chief of Naval Reserve.
3. He or she will forward appropriate sailing documentation to the Chief of Naval Reserve within 30 days after the end of each complete year following acceptance of the commission.
4. He or she will perform at least two weeks of active duty for training during each complete year following acceptance of the commission unless the requirement is waived by the Chief of Naval Reserve.
5. He or she will have no monthly participation requirement while a member of the Merchant Marine Reserve, U.S. Naval Reserve, but may enroll in and complete Navy correspondence courses, complete special assignments, or participate with designated Naval Reserve units to earn retirement credit.

Midshipmen become familiar with sailing on many different types of vessels during the program at the Merchant Marine Academy.

As in every other aspect of attendance at a service academy, it is to your best advantage to get as much information as possible on which to base your decisions. Before you sign any commitment, find out all you can about active duty and reserve obligations. Make a list of specific questions to ask the Academy representatives. For specific or general information, write to the Admissions Office, US Merchant Marine Academy, Kings Point, NY 11024–1699.

Midshipman Life

Regimental life at the Academy is a vital part of a midshipman's total educational experience, and all midshipmen are required to meet high standards of conduct and discipline. The regimental program is designed to provide the midshipmen with leadership training and experience and to develop in them qualities of self-discipline and responsibility for effective citizenship and careers as officers and leaders in the maritime industry.

Although midshipmen devote the greatest amount of time to classes, laboratories, seminars, and study, the Regimental System also makes demands on their time. These two areas of life—the academic and the regimental—are highly compatible and together produce the type of graduate for which the Academy is noted.

Fundamental to the regimental program is the Class System of responsibilities, duties, and privileges. First Classmen (seniors), under the supervision of the Commandant of Midshipmen and his staff, exercise command of the Regiment of Midshipmen. All First Classmen and some Second Classmen have opportunities to serve in Midshipman Officer billets. The Regiment of Midshipmen, under the overall command of the Midshipman Regimental Commander, is divided into three battalions, each under the command of a Midshipman Battalion Commander. These top-ranking midshipman officers work closely with the Commandant of Midshipmen in formulating and carrying out policies and procedures. They thereby receive practical leadership experience that helps to develop self-confidence, improves their understanding of human relations, and instills in them a sense of responsibility. First Classmen are granted privileges commensurate with their seniority and responsibility.

The juniors and sophomores, called Second Classmen and Third Classmen, are primarily responsible for assisting the First Classmen in the indoctrination and orientation of the freshmen, called plebes. The upperclassmen ensure that plebes observe proper military bearing and etiquette and indoctrinate them in the history and traditions of the Academy. The upperclassmen also participate in three programs designed to provide leadership training and an expanded role in the day-to-day administration of the Regiment and the Class System: the adjutant and squad leader programs for the Second Class, and the big brother/sister program for the Third Class. The privileges granted the Second and Third Class are less than those enjoyed by the First Class but more liberal than those granted to plebes.

The Class System requires that all plebes conform to exacting standards of conduct and bearing to facilitate their transformation to responsible members of the Regiment. These standards help the candidates to develop self-discipline, mental alertness, and correct military bearing.

The Fourth Class year under the Regimental and Class systems is a tough period of training and conditioning for both a life at sea and the many other challenges that face midshipmen during their training and as graduates. Regimental life at Kings Point is a primary reason why graduates of the Academy are so highly valued by all segments of the maritime industry and the armed forces for their bearing, maturity, and ability to get the job done.

The Honor Concept belongs to the Regiment of Midshipmen: it is contained in one sentence: "A midshipman will not lie, cheat, or steal." This one sentence must be accepted and supported by every member of the Regiment, and every member of the Regiment must accept the consequences of any violation of the Honor Concept. Midshipmen must understand that the Honor Concept is in effect to protect them in their daily living, to give greater value to their degrees, and to instill in them the principles of honesty and integrity that are so essential to a full and rewarding life.

Plebe Indoctrination Program

The newly appointed plebe reports to the Academy in early July for two weeks of indoctrination prior to the beginning of classes. During the indoctrination program and most of the remainder of plebe year, the new midshipmen undergo an intensive program of regimental training and indoctrination. One of the most important abilities learned at the Academy is to value and budget one's time. The daily schedule for midshipmen is rigid, requiring them to accomplish tasks in limited spans of time as well as to account for themselves at all times when on Academy grounds. The life of a plebe is demanding, but the new midshipmen soon become well versed in Academy traditions and develop a keen sense of pride and esprit de corps.

A typical day during indoctrination is as follows:

0540	Reveille—all hands
0545	Morning calisthenics
0610	Showers; clean rooms; cleaning stations
0645	Morning mess
0715	All hands clear Mess Hall; cleaning stations
0735	Morning inspection
0750	Call to colors
0800	Morning colors
0810–1200	Four Regimental periods of class
1215	Noon mess formation
1310–1600	Three Regimental periods of class
1600–1800	Intramurals; free time
1800	Showers
1830	Evening mess

1900–2100	Evening activities
2100	Showers
2130	Tattoo; accountability check
2140	Taps
2145	Lights out

Once the indoctrination program has been successfully completed, Academy life has a normal routine. A midshipman's day begins with reveille, 0610 for plebes and 0710 for upperclassmen. Breakfast, optional for upperclassmen, is at 0640. Following colors formation at 0800, midshipmen proceed to class, and classes continue until noon. Lunch is at 1220, and afternoon classes are scheduled from 1310 until 1700 (5 p.m.). Since some class periods are assigned as study hours, a midshipman spends about five hours in class each weekday. The hours between 1700 and 1900 are free time and are normally devoted to varsity athletics, intramurals, club meetings, extra study, or some form of extracurricular recreation. Following the evening meal, study period extends until the evening accountability check at 2310 (11:10 p.m.). Midshipmen may continue to study in their rooms after 2310; taps is sounded at 2200 (10 p.m.). Saturday mornings are devoted to Regimental parades and inspections, but the remainder of each weekend is used for liberty and recreation.

All midshipmen receive about two weeks of leave at Christmas and four days at Thanksgiving and Easter. In addition, they receive annual leave during the month of July. Upperclassmen normally are granted weekend liberty from 1230 on Saturday until 2100 on Sunday. Plebes are not granted overnight liberty until well into the year. The Commandant may grant sick leave or emergency leave and may also grant special leave or liberty for extracurricular activities and special events.

The Academy seeks to promote the growth of each midshipman as a whole person and is concerned with physical development as well as the development of character and intellect. Physical fitness and athletics are therefore an important part of Academy life. The varsity athletic program is comprehensive, but emphasis on intercollegiate competition is kept in harmony with the academic obligations of the midshipmen. The Academy fields varsity teams in twenty-two sports: baseball, basketball, bowling, crew, cross-country, golf, rifle, pistol, sailing, soccer, volleyball, swimming, tennis, indoor track, outdoor track, football, and wrestling. Women participate on varsity swimming and sailing teams and field a varsity volleyball team and three clubs: basketball, crew, and softball.

The intramural athletic program is organized around twenty activities and is scheduled during the recreational hours. Competition is conducted on individual, dual, and team bases, enabling each person to choose the type of activity best suited to his or her interests and capabilities. The intramural program includes: basketball, bowling, cross-county, handball, rowing, sailing, softball, swimming, table tennis, touch football, track and field, volleyball, wrestling, soccer, paddleball, rifle, pistol, badminton, and billiards. Club sports are offered for men in rugby, hockey, lacrosse, martial arts, and weight lifting.

Under the guidance of a professional sailing master, the Kings Point Sailing Squadron offers midshipmen an opportunity to enjoy top competitive intercollegiate and ocean racing almost year round. The Academy's fleet of twenty 420's, twenty Tech dinghies, and twenty Lasers represents one of the finest collegiate facilities in the country. Daily practices and nearly eighty regattas per year enable the Academy to field one of the nation's top ten nationally ranked intercollegiate teams.

The Academy Counseling Program is directed toward providing each midshipman with the counseling required to resolve problems, explore career patterns, and ensure an optimum educational experience. Particular attention is focused on the plebes, for whom adjustment to the Academy and the collegiate environment is the most difficult. Academic counseling is provided by faculty advisers. Remedial studies and special tutoring are provided by upperclassmen supported by faculty members. Personal counseling is provided by the Medical Department, officers of the Office of External Affairs, including those of the Admissions Office, and the chaplains and officers of the Commandant's staff.

More than forty clubs and extracurricular activities are available to midshipmen. These range from musical, literary, debate, and camera activities to chess, scuba diving, and international relations. A full-time social hostess coordinates activities and assists with the planning of formal Regimental dances, frequent informal mixers, and an assortment of other events designed to enrich the social life of the midshipmen. Participation in religious activities and attendance at chapel services are voluntary. The Chapel serves all faiths.

The major cost of attendance at the Academy is borne by the government. Midshipmen are provided with comfortable quarters and well-balanced meals. Medical and dental care are provided at the Academy Infirmary. Initial issuances of uniforms and textbooks are provided by the government at no charge to the midshipman. To assure uniformity of appearance, quality, and cost, all required items are purchased by Academy staff to specifications approved by the Commandant of Midshipmen or the Academic Dean. If additional textbooks are considered desirable, it is the responsibility of the midshipman to pay for them. Each midshipman is required to purchase or possess an electronic hand calculator.

Midshipmen are required to pay for mandatory educational supplies not provided by the government, midshipman activity fees (athletics, cultural events, health service, student newspapers, yearbook, etc.), and Personal Service Fee (self-service laundry, barbershop, and tailoring). Deposits for these mandatory fees range from $400 to $700 and must be paid prior to reporting to the Academy each year. During shipboard training, midshipmen are provided with quarters, meals, and medical care and are paid about $500 per month by the commercial companies, less certain minor expenses.

Conclusion

The culmination of all the hard work that goes into successfully completing the rigorous four-year programs at the service academies is the awarding of a commission as a Second Lieutenant or Ensign in the service of the United States. The commissioning ceremony is both a solemn and a joyous occassion. It is a proud moment in the life of dedicated young men and women, and the ceremony is usually attended by friends and parents. The newly commissioned officers take an oath of office in which they pledge faithful service to the United States.

The Oath of Office

> I, (first name, middle name, last name and service number), having been appointed an officer in the Army (Navy, Air Force, Marine Corps) of the United States, in the grade of Second Lieutenant (Ensign), do solemnly swear (or affirm) that I will support and defend the Constitution of the United States against all enemies, foreign and domestic; that I take this obligation freely, without any mental reservation or purpose of evasion; and that I will well and faithfully discharge the duties of the office upon which I am about to enter, so help me God.

This oath must be understood by all officers; it is a personal statement, made freely without qualification, that the officer will conscientiously and faithfully serve the United States of America.

Today, the armed forces have a tremendous need for well-educated men and women from all disciplines—military, scientific, technical, and humanistic. The economic, scientific, political, cultural, and social implications of high-level military decisions that are made almost daily must be studied and evaluated by the most knowledgeable minds in the United States. These decisions are made by men and women wearing the uniforms of the armed forces. This analytical and creative thinking is responsive to the education, training, experience, and self-discipline acquired through years of dedicated service.

That is the challenge facing young men and women of this country today—the opportunity to serve the nation as professional officers in the Armed Forces. The challenge can only become more difficult in the years ahead as the United States and the rest of the world deal with increasingly difficult problems of resource shortages, overpopulation, nuclear arms races, space exploration, and complicated international relations. Qualified service academy graduates go on active duty as well-trained junior officers who have an opportunity to affect the future. These young men and women are an important part of the future.

To be productive and effective, every man and woman needs to feel job satisfaction in his or her chosen career field. A "job" in this sense does not mean being paid for working at a task from nine to five and then rushing home to forget about it. Men and women serving as officers in the Armed Forces feel real satisfaction and fulfillment in accomplishing their work. A service career is an occupation and a profession in which the officer has total involvement. It is also a demanding career that calls for sacrifice, family separation at times, and a desire to serve one's country. Fortunately, many young men and women in the United States have high ideals and a deep sense of patriotism. In talking with cadets all over the country, I was greatly impressed with their seriousness of purpose and sense of values. They expressed a desire to contribute to the security of the United States and to render important service to society. It is not a high-salaried job and material advantages that they are seeking; rather, it is the opportunity to serve a useful purpose in the world. Their high aspirations are compelling, and the career that they finally settle on must satisfy both their intellectual and emotional requirements and the vital human need of fulfillment of life through work that is worthy of a person's full effort.

For many of these young men and women, the answer may be found in a career in the Armed Forces. It is a life of service and dedication to country, and one of traditional honor and prestige. Few fields hold greater promise for an ambitious young person seeking immediate responsibility, and no other field so combines adventure and security. There are opportunities for education and growth, travel and inspiring experiences. Officers know that their duties are vital to the security of the nation, and they welcome the growing responsibilities for the common welfare that come as they mature. It is a demanding life of challenge and change, unlimited opportunities, and high goals, and it brings rich rewards to the men and women who dedicate themselves and their abilities to it. Our world is not so much inherited from our ancestors as it is on loan from our children. Our Armed Forces insure that we are able to repay that loan.

Appendix

On the following pages are points of contact and sample letters for congressional nomination.

Points of Contact

UNITED STATES MILITARY ACADEMY
Admissions Office
West Point, NY 10996-1797
(914) 938-4041

UNITED STATES NAVAL ACADEMY
USNA Candidate Guidance Office
Annapolis, MD 21402-5019
(202) 267-4361 or (800) 638-9156

UNITED STATES AIR FORCE ACADEMY
Admissions Office
Colorado Springs, CO 80840-5651
(719) 472-2520

UNITED STATES COAST GUARD ACADEMY
Admissions Office
New London, CT 06320-4195
(203) 444-8444

UNITED STATES MERCHANT MARINE ACADEMY
Admissions Office

Kings Point, NY 11024-1699
(516) 773-5391 or (800) 732-6267 (outside NY state)

UNITED STATES MILITARY ACADEMY

Request for Congressional Nomination

Date ____________________

The Honorable ____________________		The Honorable ____________________
United States Senate	OR	House of Representatives
Washington, D.C. 20510-0001		Washington, D.C. 20515-0001

Dear Senator ____________________ Dear Mr./Ms. ____________________

I desire to attend the United States Military Academy and to be commissioned in the Regular Army. I respectfully request that I be considered as one of your nominees for the class entering West Point in June 19–.

The following data are furnished for your information:

Name: ____________________

Permanent Address: ____________________

Telephone Number: ____________________

Temporary Address and telephone number (if different from preceding):

____

____

Date of Birth: ____________________

High School: ____________________

Social Security Number: ____________________

Names of Parents: ____________________

I have/have not requested that a precandidate file be initiated for me at the West Point Admissions Office.

Sincerely,

Request for Service-Connected Nomination

Date ____________________

Director of Admissions
United States Military Academy
West Point, New York, 10996-1797

Dear Sir:

I request a nomination under the ______________________________ category for the class entering the United States Military Academy in June 19–, and I submit the following data:

Name of Applicant: ____

Address: ____
Street City State Zip Code

Telephone Number (__________) ____

Date of Birth: ____

Social Security Number: ____

Names of Parents: ____

Military Rank of Parent: ____

Social Security Number of Parent: ____

Component and Branch of Service of Parent: ______________________________________

A stepparent's military service is applicable for this category of nomination only if the applicant was adopted prior to his or her fifteenth birthday.

If parent is on active duty, you must furnish a statement of service signed by the unit adjutant.

If parent is retired or deceased, you must furnish copy of retirement order or casualty report.

Include a brief statement concerning the date, place and cause of death or the details of disability together with the claim number assigned to the veteran parent's case by the Veterans Administration (if appropriate).

Applicants for Reserve Component or Regular Army nominations should include an official statement, signed by your unit commander, providing your military status and inclusive dates of military service in accordance with AR 351-12.

Include a brief statement of the date and circumstances of the award of the Medal of Honor (if appropriate).

(ENLISTED APPLICANTS ARE REFERRED TO AR 351-12.)

Sincerely,

UNITED STATES NAVAL ACADEMY

Request for Congressional or Vice Presidential Nomination

The Vice President
The White House
Washington, DC 20501

The Honorable ______
House of Representatives
Washington, DC 20515

The Honorable ______
United States Senate
Washington, DC 20510

Dear Mr. Vice President or Senator __________ or Mr. ______________________:
(Senate) *(House of Representatives)*

It is my desire to attend the United States Naval Academy. I respectfully request that I be considered as one of your nominees for the class entering in July 19____.

The following personal data are provided for your information:

Full name: ____
(Print as recorded on birth certificate)

Name of parents: ____

Address: *(Use zip code and provide phone number)*

Permanent ______________________ Mailing ______________________

______________________ ______________________

Telephone ______________________ Telephone ______________________

Date of birth: ____________________ Place of birth: ____________________

Social Security number: ____

High school attended: ____
(Name and address)

Date of high school graduation: ____________________ Sex: ______________

My approximate standing is ______________ in a class of ____________________

I have/have not sent a Precandidate Questionnaire to the Naval Academy.

I have requested my high school transcript of work completed to date to be forwarded to your office as soon as possible. I have also listed on the reverse side the results of any ACT or College Board test scores that I have taken.

I have been active in high school extracurricular activities as indicated on the reverse side. I should greatly appreciate your consideration of my request for one of your nominations.

Sincerely yours,
(Signature)

Request for Presidential Nomination

To: Nominations and Appointments Officer, United States Naval Academy, Annapolis, MD 21402-5018.

Dear Sir: Date: ____________

I request a Presidential nomination to the United States Naval Academy for the class which will enter in July 19___ and submit the following data:

Name: ____________
(Give full name as shown on birth certificate or, if changed, attach copy of court order.)
Address: *(Use zip code and provide phone number)*
Permanent ____________ Mailing ____________

____________ ____________

Phone ____________ Phone ____________

Date of birth: ____________

Social Security number *(must be filled in)*: ____________

Name and address of high school/college: ____________

Month/year of graduation: ____________ Sex: ____________

Ethnic origin: ____________ *Black, Oriental, Hispanic, native American (American Indian and native Alaskan), Puerto Rican, Caucasian, etc.*

Congressional district & state: ____________

Applying to Congressmen *(names)*: ____________

Highest scores: PSAT V ___, M ___; SAT V ___, M ___;
ACT E ___, M ___

Uncorrected vision: Right 20/___, Left 20/___, Corrected R 20/___, L 20/___

If member of military, check here ___. List rank, serial number, component, branch of service, and organizational address on reverse side of this form.

Information concerning parents' military service:

Name of parent/s: ____________
(Parent's rank, serial number, component, and branch of service; if parent is retired with pay please indicate)

Sincerely yours,
(Signature)

UNITED STATES AIR FORCE ACADEMY

Request for Congressional or Vice-Presidential Nomination

Date

Honorable (Name of Appropriate Authority)
House of Representatives OR United States Senate
Washington, DC 20515 OR Washington, DC 20510

OR

The Vice President
Old Executive Office Bldg
Washington DC 20501

Dear Mr./Mrs./Ms. (Name) OR Dear Senator (Name) OR Mr. Vice President

I want to attend the Air Force Academy and to serve in the United States Air Force. I request that I be considered as one of your nominees for the class that enters the Academy in June 19—.

My pertinent data is:
Name (print as recorded on birth certificate):
Social Security Number:
Permanent address (street, city, county, state zip code):

Temporary address (if applicable):

Permanent phone number and area code:
Temporary phone number and area code (if applicable):
Name of father:
Name of mother:
Date and place of birth (spell out month):
Name and address of high school:
Date of graduation:

Approximate grade point average (GPA), rank-in-class and PSAT, ATP (SAT) and ACT if you have taken these tests. Include verbal and math scores for the PSAT and ATP (SAT) tests, and English, math, social studies, and natural science scores for the ACT test.
Extracurricular activities:

Reasons for wanting to enter the Air Force Academy:

Thank you for considering me as one of your nominees to the Air Force Academy.

Sincerely

Signature

Request for Military-Affiliated Nomination
(Use this format for any of these categories: Presidential,
Children of Deceased or Disabled Veterans,
or Children of Medal of Honor Recipients.)

Date

Director of Admissions
HQ USAFA/RR
USAF Academy
Colorado Springs CO 80840-5651

Dear Sir:

I want to attend the Air Force Academy and to serve in the United States Air Force. I request a nomination under the (name of appropriate category) for the class that enters the Academy in June 19—.

My pertinent data is:
Name (print name exactly as it appears on the birth certificate or if legally changed, attach a copy of the court order):
Social Security Number:
Permanent address (street, city, county, state, zip code):

Temporary address (if applicable):

Permanent phone number and area code:
Temporary phone number and area code (if applicable):
Date and place of birth (spell out month):

If member of military include rank regular or reserve component, branch of service and organizational address including PSC Box Number:

If previous candidate, indicate year:

Information on parents:

Name, rank, social security number, component and branch of service:
Parent active duty (attach statement of service dated and signed by current personnel officer specifying all periods of active duty and any breaks therein).
Parent retired or deceased (attach copy of retirement orders or casualty report; include Veterans Administration (VA) claim number and VA office where case is filed, if appropriate; include brief statement with date and circumstances of Medal of Honor award, if appropriate).

Sincerely,

Signature

U.S. MERCHANT MARINE ACADEMY

Request for Congressional Nomination

Date ________________

The Honorable ________________ or The Honorable ________________
House of Representatives United States Senate
Washington, D.C. 20515 Washington, D.C. 20510

Dear ________________:

It is my desire to attend the United States Merchant Marine Academy. I respectfully request that I be considered one of your nominees for the class entering the Academy in the summer of 19__.

The following personal data are provided for your information:

Full name ________________________________
(Printed as recorded on birth certificate.)

Name of parents ________________________________
Address (include ZIP code and phone number)

Permanent Temporary

________________ ________________

________________ ________________

________________ ________________

My date of birth: ________________ Place of birth: ________________

Social Security number: ________________________________

High school attended: ________________________________
(Name and address)

My approximate standing is ________________ in a class of ________________

I have requested that a high school transcript of my work completed to date be forwarded to your office as soon as possible. I have also listed on the reverse side of this letter my results on the ACT or College Board examinations.

I have been active in high school extracurricular activities as indicated on the reverse side.

I would greatly appreciate your consideration of my request for one of your nominations.

Sincerely yours,
(Signature)

Index